土人景观

北京土人景观与建筑规划设计研究院／编

Turenscape

最新规划设计表现图集

Compilation of Turenscape's recent design and planning projects

中国林业出版社

China Forestry Publishing House

目录

CONTENTS

01风景区景观规划与设计　03公园区景观规划与设计　05校园与科技园景观规划与设计

02滨水区景观规划与设计　04商业区与工业区景观规划与设计

01风景区景观规划与设计 01 Landscape Planning and Design of Scenic Areas

004	北京市雁栖湖生态发展示范区整体规划	Planning of Yanxi Lake Eco-development Demonstration District, Beijing
008	秦皇岛圆明山山野森林公园	Yuanming Mountain Wild Forest Park, Qinhuangdao
010	上汤大佛景区核心区景观及建筑设计	Landscape and Architecture Design of Core Area of the Grand Buddha, Shangtang
014	《烟雨凤凰》大兴山水实景演出剧场及周边配套区域	Live-performance Theatre and Surrounding Supporting Area of the Show of Yan Yu Feng Huang
018	仙姑岭风景旅游区项目策划与概念规划设计	Project Theme and Concept Planning Design of Xianguling Scenic Area
020	中国羊山古镇国际军事旅游度假区规划及建筑设计	Planning and Architecture Design of Yangshan Ancient Town International Military Tourist Resorts, China
024	秦岭悠然山风景区核心区景观建筑方案设计	Landscape Architecture and Architecture Design of Core Area in Qinling Youran Mountain Scenic Area
030	千岛湖金峰生态度假村总体规划	Master Planning of Qiandao Lake Jinfeng Eco Resort

02滨水区景观规划与设计 02 Landscape Planning and Design of Riverfront

036	美国明尼阿波利斯密西西比河滨水设计	Mississippi Riverfront Design in Minneapolis, USA
044	秦皇岛海岸线重点区域景观设计及建筑设计	Landscape Architecture and Architecture Design of Coastal Key Areas in Qinhuangdao
046	秦皇岛市红旗北路西侧绿地公园	Greenspace Park at West Side of Hongqi North Road in Qinhuangdao
048	鹤壁市淇河生态公园景观工程	Landscape Engineering of Qihe Eco Park in Hebi
050	连云港东河新城城市设计	Urban Design of Donghe New Town in Lianyungang
054	大连小窑湾景观设计	Landscape Architecture of Xiaoyaowan, Dalian
058	烟台市区东北部岸线景观修建性详细规划	Constructive Detailed Planning of Northeastern Coastline Landscape of Yantai Downtown
060	大同市十里河生态廊道	Shilihe Eco Corridor in Datong
062	上海后滩公园	Houtan Park, Shanghai
066	天津北塘地区北区景观设计	Landscape Architecture of North Tanggu District, Tianjin
068	天津海河教育园区（北洋园）一期	Phase 1 of Haihe Education Area (Beijing Garden) in Tianjin
070	天津市东丽湖滨水景观规划及会议半岛详细城市设计	Landscape Architecture of Dongli Lake Waterfront and Detailed Urban Design of Huiyi Peninsula
072	浙江省宁波梅山保税港区中央运动公园城市设计	Urban Design of Central Activity Park of Meishan Bonded Port in Ningbo, Zhejiang
074	浙江三门滨海新城横港及金鳞湖景观设计	Landscape Architecture of Coastal New Town Heng Port and Jilin Lake in Sanmen, Zhejiang Province

03公园区景观规划与设计 03 Landscape Planning and Design of Park

076	芝加哥北格兰特公园	North Grant Park in Chicago
079	宜昌市运河公园	Channel Park in Yichang
082	巴西里约热内卢奥林匹克公园	Olympic Park in Rio, Brazil
086	吉林长春南关区光明公园景观工程	Landscape Engineering of Guangming Park in Nanguan District of Changchun, Jilin
088	山西大同晋华宫矿矿山公园概念规划	Concept Planning of Mine Park at Jinhuagong Mine in Datong, Shanxi
092	西班牙马德里公园	Madrid Park, Spain
094	哈尔滨群力国家城市湿地公园	Qunli National Urban Wetland, Harbin
098	苏州高新区真山公园	Zhenshan Park in High-Tech District, Suzhou

04商业区与工业区景观规划与设计 04 Landscape Planning and Design in Commercial and Industrial Areas

102	郑州高新区彩虹花园景观改造工程	Landscape Transforming Project of Rainbow Garden in Zhengzhou High & New Technology Industries Development Zone
104	睢宁县新市街景观与商业建筑改造	Landscape and Commercial Architecture Transforming of Xinshi Street in Suining
106	大同煤气厂景观改造工程	Landscape Transforming Project of Datong Gas Plant
110	中国宣纸传习基地修建性详细规划	Constructive Detail Planning of Rice Paper Pass and Learn Base

05校园与科技园景观规划与设计 05 Landscape Planning and Design of Campus and Science Park

114	北京中关村科技园区昌平园研发基地城市设计方案	Urban Design Scheme of Research & Development Base in Changping Sub-park, Zhongguancun Science Park
118	承德市宽城民族教育园区总体规划	General Planning of National Education Park in Town of Kuancheng, Chengde City
122	秦皇岛市党校建筑初步设计、景观工程设计	Preliminary Design, Landscape Engineering Project of Party School in the City of Qinhuangdao
128	洛阳师范学院校园总体规划方案设计	General Planning Scheme of the Campus of Luoyang Normal University
132	辽宁公安司法管理干部学院景观设计	Liaoning Administration College of Police and Justice
136	云南红河技师学院概念规划及修建性详细规划	Conceptual and Constructive Detail Planning of Honghe Technician Institute, Yunan Province
140	昆明理工大学呈贡校区规划设计	Planning Design of Chenggong Campus, Kunming University of Science and Technology
146	宜春学院高安校区建筑工程设计	Architectural Engineering Design of Gao'an Campus, the College of Yichun

06公共建筑设计 06 Public Architecture Design

150	林州红旗渠博物院建筑工程设计	Architectural Engineering Design of Red Flag Canal Museum, Linzhou
152	江苏省睢宁县地藏寺建筑概念及周边景观设计方案	Architectural Concept and Surrounding Landscape Architecture Scheme of Dizang Temple, County of Suining, Jiangsu Province
158	九江县文化艺术中心建筑工程设计	Architectural Engineering Design of Culture and Art Center, County of Jiujiang
160	北京环渤海高端总部基地	High-end Headquarters Base in the Bohai Sea, Beijing
161	开远市体育中心建筑工程设计	Architectural Engineering Design of Sports Center, Kaiyuan City
162	韶山藏书楼项目规划设计	Planning Design of Shaoshan Library Scheme

07城市景观规划与设计 07 Urban Design and Planning

164	六盘水市中心区荷城古镇改造及周边用地修建性说细规划	Redevelopment of Old Dutch Town, Liupanshui CBD, and rezoning and masterplanning of surrounding area
170	海南屯昌城市中心区城市设计	Urban Design of Tunchang city centre, Hainan
174	山海关城区总体城市设计	Urban Design of Shanhaiguan Metropolitan Area
176	武汉市中国光谷依托邦城市设计	Urban Design of China Optics Valley, Wuhan City
182	株洲市轨道科技城城市景观风貌规划暨重点区域城市设计	Urban Design of key zones and landscape Design and planning of Track Technology Park, Zhuzhou City
184	重庆市江津区北部新城城市设计	Urban Design of northern sector, Jiangjin District, Chongqing
188	重庆酉阳新城城市设计	Urban Design of new Youyang City, Chongqing
192	江苏高淳科技创业特别社区城市设计	Urban Design of Innovation Technology Special Zone, Jiangsu, Gaochun
196	陕西省西咸新区渭河生态景观带总体方案设计	Ecological landscape framework Design for Wei River, Xixian New Area, Shanxi
198	大英县郪江新城城市设计	Urban Design for New Quarter Qijiang City, Daying
204	新疆鄯善县城区重点地段城市修建性详细规划	Masterplanning of key urban areas in Shanshan, Xinjiang
210	遵义云顶阳光项目修建性详细规划设计	Masterplan for Wenting Project, Zunyi, Guizhou

08动画表现 08 Animation movie

214	美国明尼阿波利斯密西西比河滨水设计	The Minneapolis Waterfront Design Concept
216	巴西里约热内卢奥林匹克公园设计	A 'Carioca' Legacy: Olympic Park Rio 2016
218	北京环渤海高端总部基地城市景观轴线景观设计	Landscape Design for Beijing Bohai-rim Advanced Business Park Landscape Axis
220	广州市天河智慧城生态基础设施暨开放空间研究及天河智慧城生态基础设施规划	Ecological Infrastructure and Open Spaces Study of Zhihui City and Ecological Infrastructure Planning for Zhihui City, Tianhe District
222	芝加哥北格兰特公园设计	The Art Field: The Chicago North Grant Park
224	五里界生态城暨中国光谷伊托邦城市设计	Wuli Eco-city and Urban Design of China Optics Valley
226	六盘水荷城古镇改造及周边用地修建性详细规划	Redevelopment of Old Dutch Town, Liupanshui CBD, and rezoning and masterplanning of surrounding area
228	台州鉴洋湖湿地公园一期修建性详细规划	Kamyang Lake Wetland Park – Stage 1 Masterplan
230	北京市雁栖湖生态发展示范区整体规划	Master plan of Yanqi Lake ecological demonstration region, Beijing
232	台儿庄古镇景观规划暨核心区域城市设计	Landscape planning and urban Design of central Taierzhuang Ancient Town
234	云南祥云"新云南驿"项目概念性总体规划	Conceptual master planning of "New Yunnan Courier Station", Xiangyun, Yunnan province
236	吉安市江子头村改造规划及建筑和景观概念设计方案	Reconstruction Planning and Conceptual architectural and landscape planning of Jiangzitou Village, Ji'an
238	山西省临汾市洪洞县滨河西区城市设计	Urban Design of West Riverside of Hongdong County, Linfen, Shanxi
240	中国滑县道口古镇	Daokou Ancient Town, Hua County
242	郑州IT产业园概念规划	Conceptual Planning of Zhengzhou IT Industrial Park
244	钟祥市明显陵风景区	The Ming Tombs Scenic Area, Zhongxiang
246	重庆市江津区新城城市设计	Urban Design of Jiangjin district of Chongqing City
248	宜昌廊桥水岸	Yichang Bridge and Water Residential Area
250	济宁阳光城市花园	Sunshine Urban Garden Residential Area, Jining
252	京林理想城	Jinglin Ideal City
254	海拉尔河东新城核心区城市设计	Urban Design of Central Hailaer East New Town
256	胶南市风河两岸景观工程	Landscape Project Feng He Riverside, Jiaonan
258	山东海阳龙栖城修建性概念规划	Construction Conceptual Planning of Longqi City, Haiyang, Shandong
260	洛阳师范学院校园总体规划方案	Master plan of Luoyang Normal University Campus

09 获奖 08 Honors

北京市雁栖湖生态发展示范区整体规划

项目地点：北京市 怀柔区
项目规模：21平方千米
项目类别：城市规划、景观设计
委 托 方：北京市规划委员会

Planning of Yanxi Lake Eco-development Demonstration District, Beijing

Location: Huairou, Beijing
Area (size): 21km^2
Project Type: Urban Planning, Landscape Architecture
Client: Beijing Municipal Bureau of Urban Planning

秦皇岛圆明山山野森林公园

项目地点： 河北省 秦皇岛市
项目规模： 20公顷
项目类别： 景观设计
委 托 方： 秦皇岛市海港区园林局

Yuanming Mountain Wild Forest Park, Qinhuangdao

Location: Qinhuangdao, Hebei Province
Area (size): $20hm^2$
Project Type: Landscape Architecture
Client: Gardening Bureau of Haigang District, Qinhuangdao

上汤大佛景区核心区景观及建筑设计

项目地点：河南省 平顶山市

项目规模：30公顷（建筑：50000平方米）

项目类别：景观设计、建筑设计

委 托 方：河南天瑞集团有限公司

Landscape and Architecture Design of Core Area of the Grand Buddha, Shangtang

Location: Pingdingshan, Henan

Area (size): $30hm^2$ (architecture: $50{,}000m^2$)

Project Type: Landscape Architecture, Architecture Design

Client: Henan Tianrui Group Co. Ltd.

Baijing

《烟雨凤凰》大兴山水实景演出剧场及周边配套区域

项目地点：湖南 凤凰县
项目规模：20公顷
项目类别：景观设计、建筑设计
委 托 方：湘西凤凰文化传播产业有限公司

Live-performance Theatre and Surrounding Supporting Area of the Show of Yan Yu Feng Huang

Location: Fenghuang, Hunan Province
Area (size): 20hm²
Project Type: Landscape Architecture, Architecture Design
Client: Xiangxi Fenghuang Cultural Industrial Co.,Ltd.

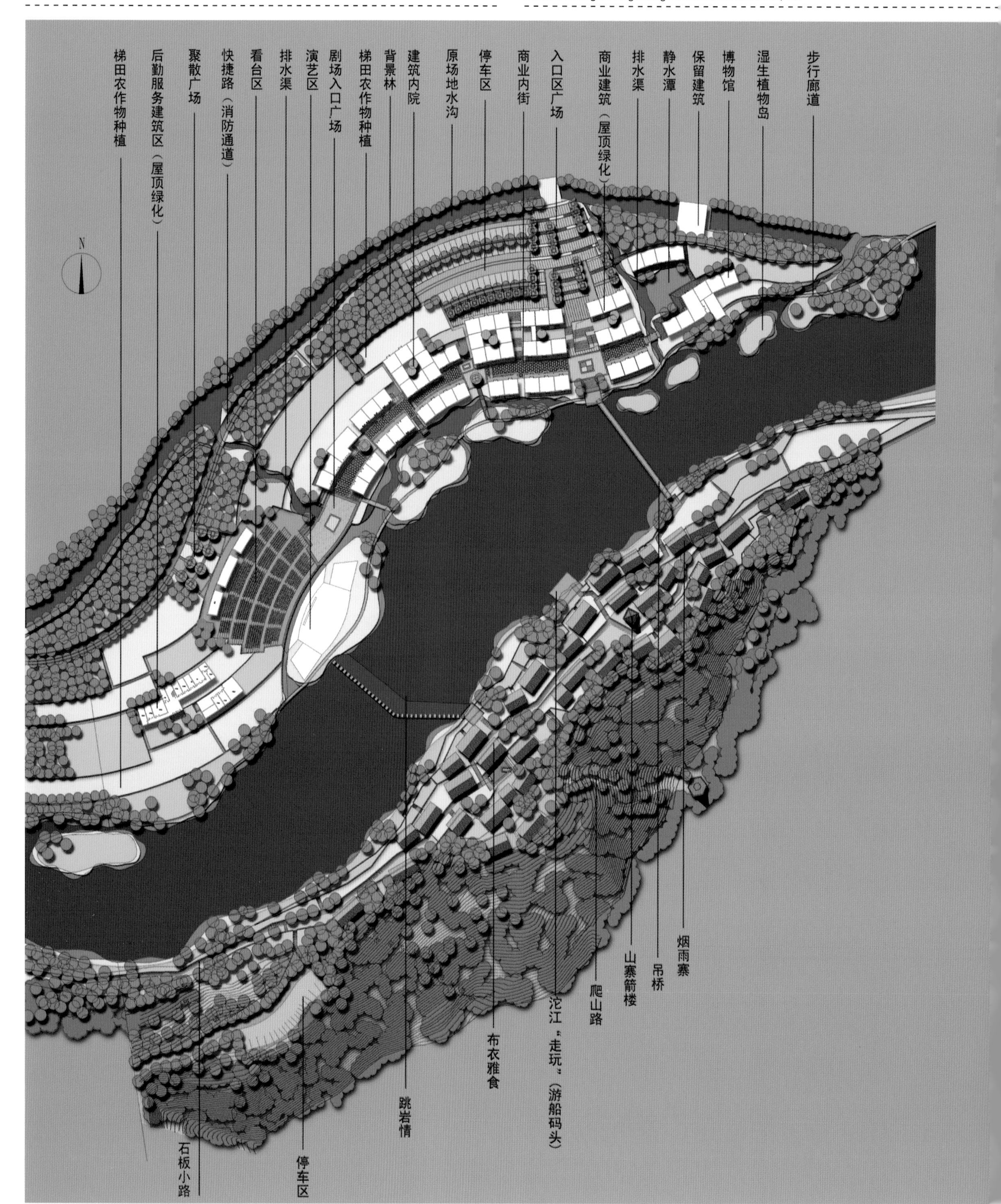

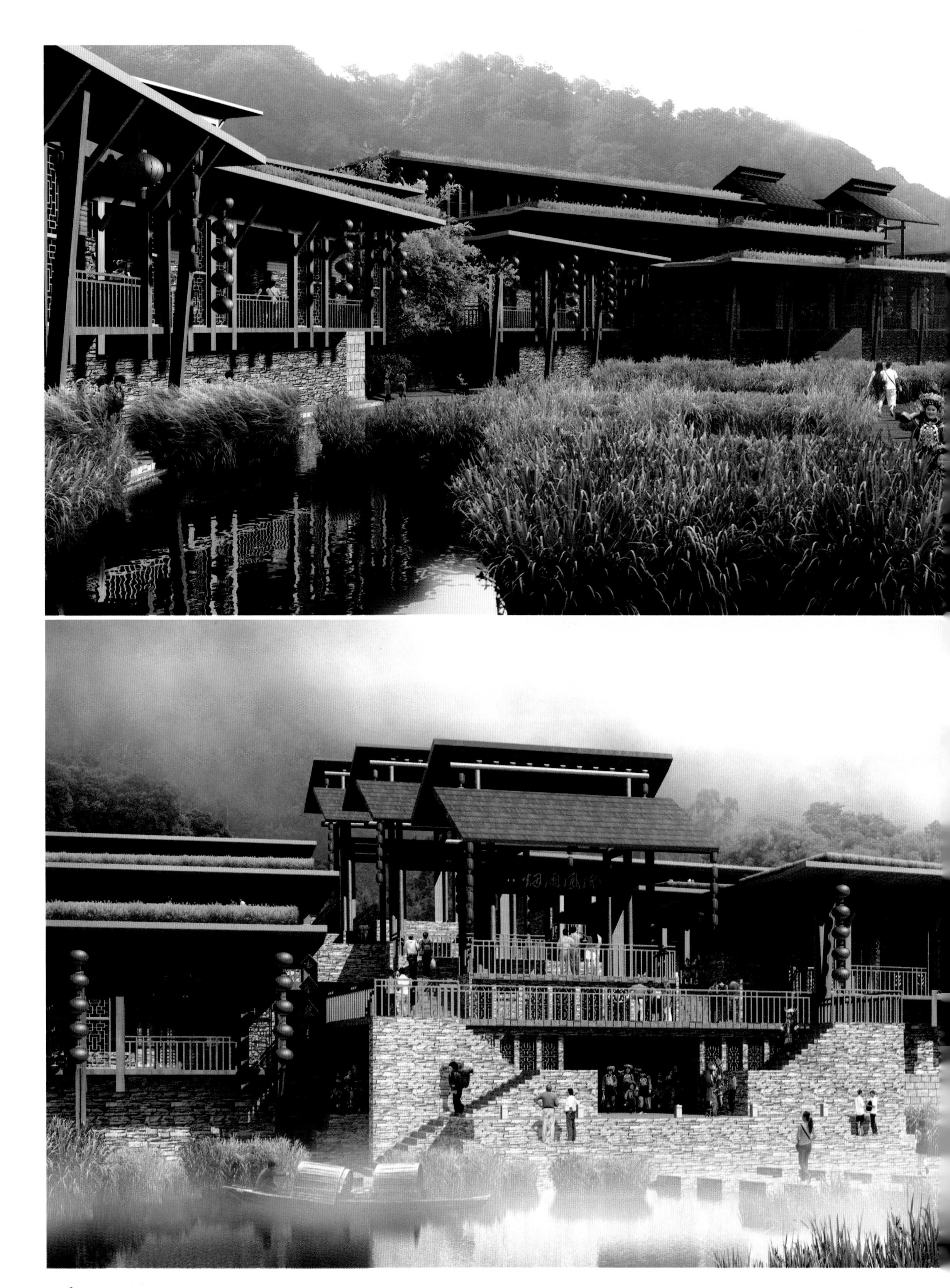

菩提書屋

仙姑岭风景旅游区项目策划与概念规划设计

项目地点：江西省 丰城市
项目规模：16.6平方千米
项目类别：城市规划
委 托 方：江西新中美实业投资有限公司

Project Theme and Concept Planning Design of Xianguling Scenic Area

Location: Fengcheng, Jiangxi Province
Area (size): 16.6km^2
Project Type: Urban Planning
Client: Jiangxi Xinzhongmei Industrial Investments Co., Ltd.

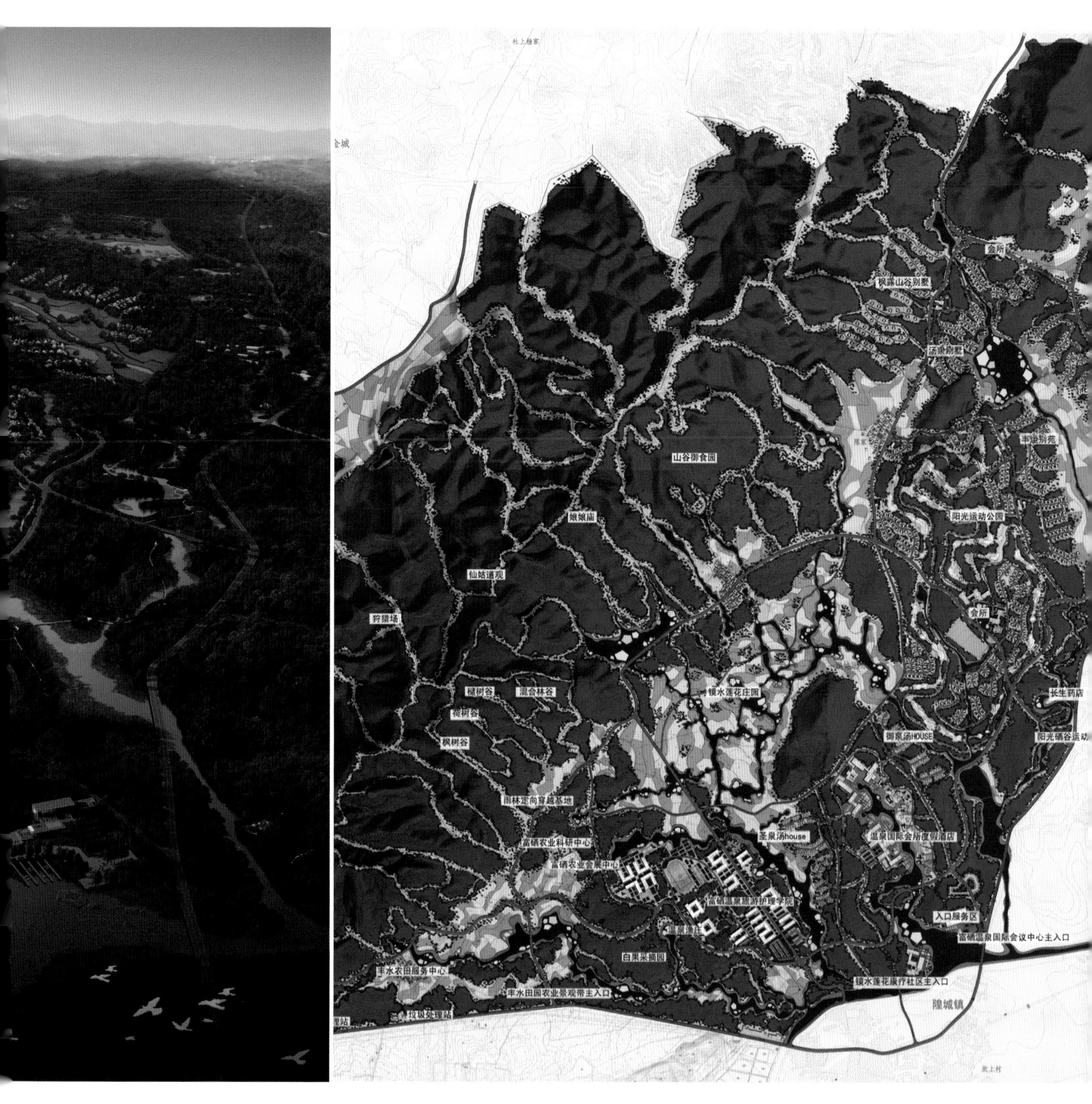

杜上杨家
会所
枫露山谷别墅
汤泉别墅
丰运别苑
山谷御食园
娘娘庙
阳光运动公园
仙姑道观
会所
狩猎场
槠树谷
混合林谷
荷树谷
镜水莲花庄园
长生药店
枫树谷
御泉汤HOUSE
阳光硒谷运动
雨林定向穿越基地
圣泉汤house
富硒农业科研中心
温泉国际会所度假酒店
富硒农业会展中心
富硒温泉旅游护理学院
入口服务区
温泉渔庄
富硒温泉国际会议中心主入口
白果采摘园
丰水农田服务中心
镜水莲花康疗社区主入口
丰水田园农业景观带主入口
垃圾处理站
隍城镇

中国羊山古镇国际军事旅游度假区规划及建筑设计

项目地点：山东省 济宁市金乡县

项目规模：15平方千米（建筑：11474平方米）

项目类别：城市规划、景观设计、建筑设计

委 托 方：金乡县人民政府

Planning and Architecture Design of Yangshan Ancient Town International Military Tourist Resorts, China

Location: 15km^2 (architecture: 11,474m^2)

Area (size): 16.6km^2

Project Type: Urban Planning, Landscape Architecture, Architecture Design

Client: Jinxiang County Council

汉文化博物馆

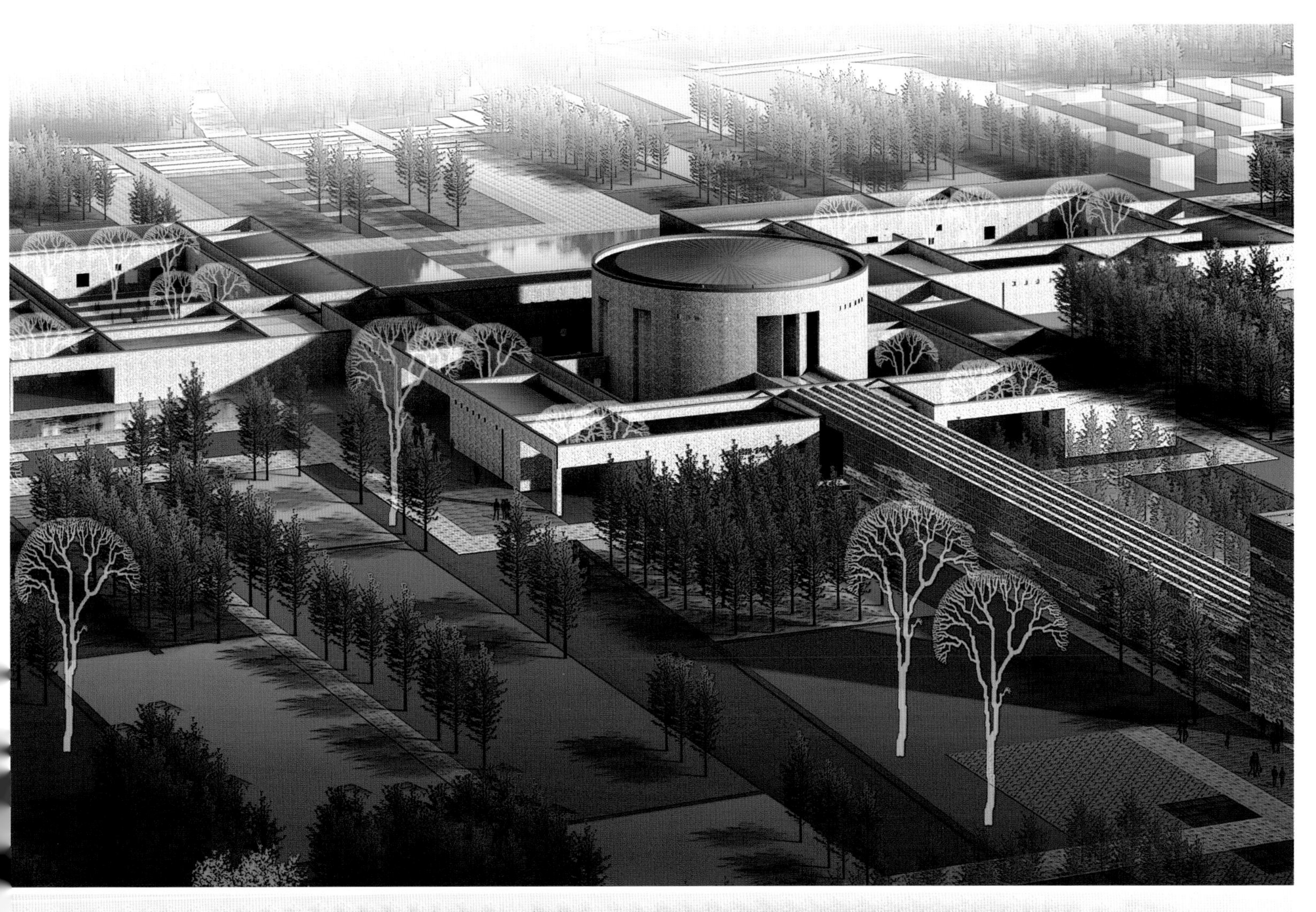

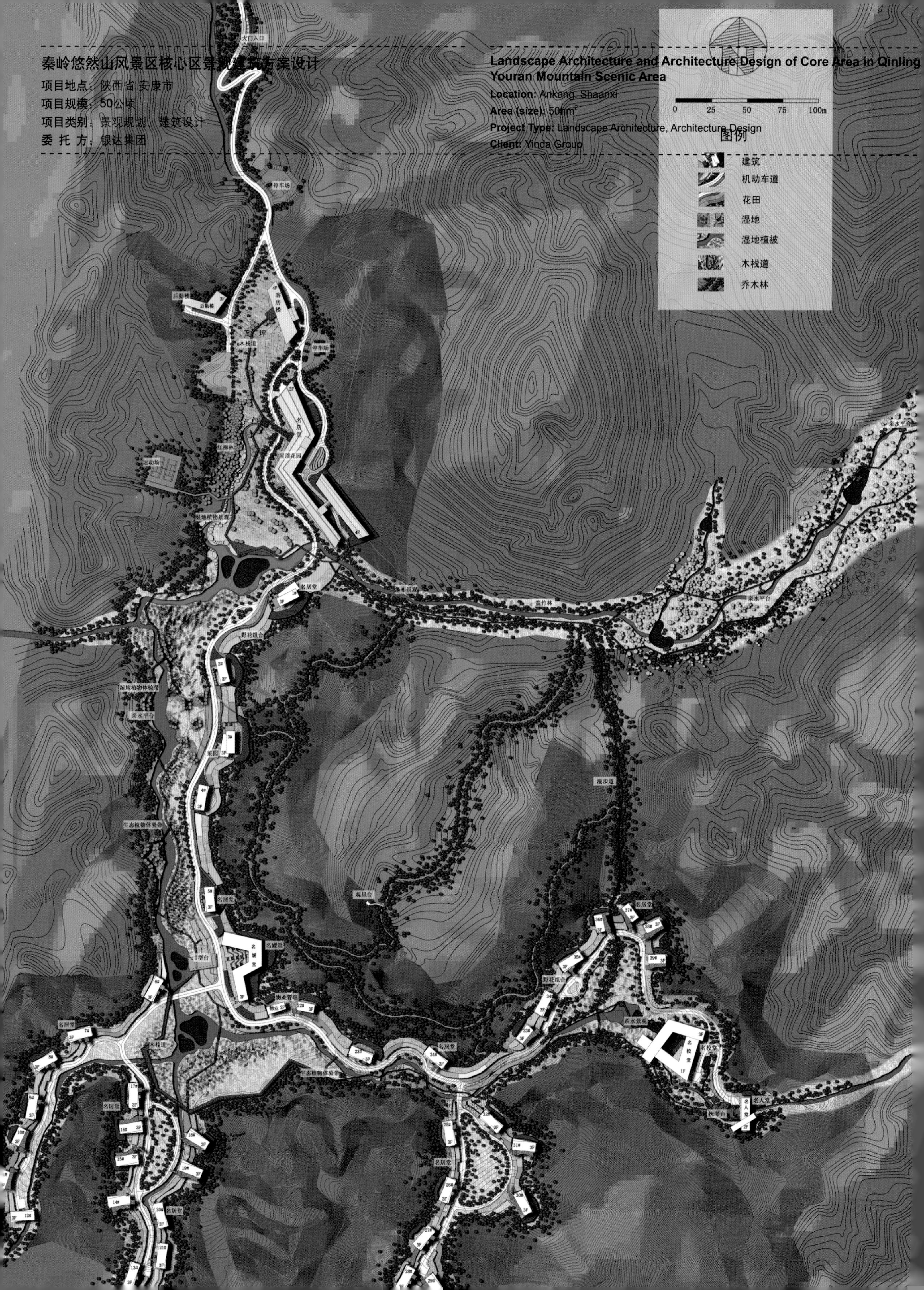

秦岭悠然山风景区核心区景观建筑方案设计
项目地点：陕西省 安康市
项目规模：50公顷
项目类别：景观规划、建筑设计
委 托 方：银达集团
Landscape Architecture and Architecture Design of Core Area in Qinling Youran Mountain Scenic Area
Location: Ankang, Shaanxi
Area (size): 50hm²
Project Type: Landscape Architecture, Architecture Design
Client: Yinda Group
0 25 50 75 100m
图例
建筑
机动车道
花田
湿地
湿地植被
木栈道
乔木林
大门入口
停车场

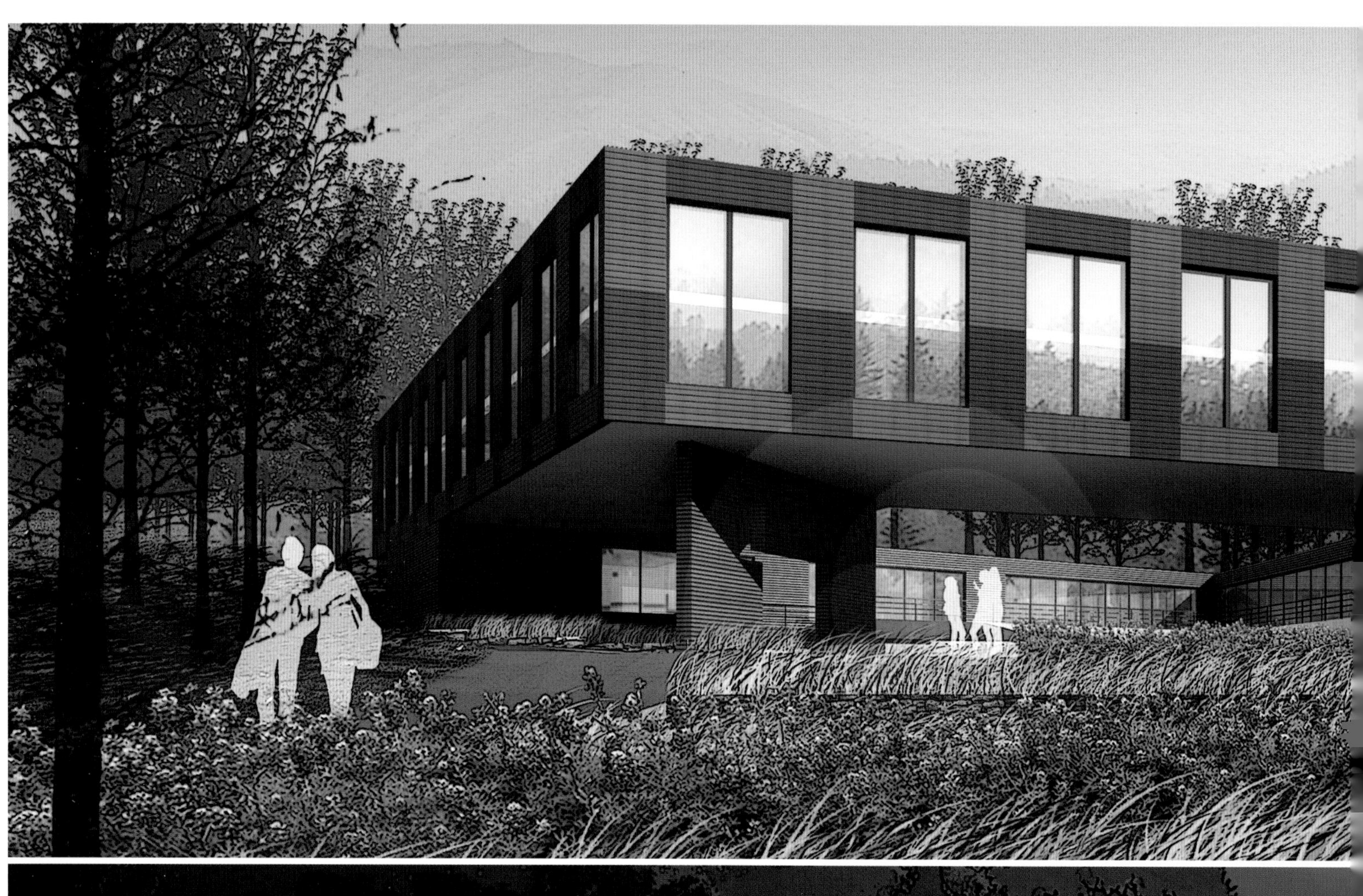

千岛湖金峰生态度假村总体规划

项目地点：浙江 杭州
项目规模：110公顷
项目类别：城市规划、建筑设计
委 托 方：杭州千岛湖生态居房地产有限公司

Master Planning of Qiandao Lake Jinfeng Eco Resort

Location: Hangzhou City, Zhejiang Province
Area (size): 110hm^2
Project Type: Urban Planning, Architecture desgin
Client: Hangzhou Qiandao Lake Eco Real Estate Co., Ltd.

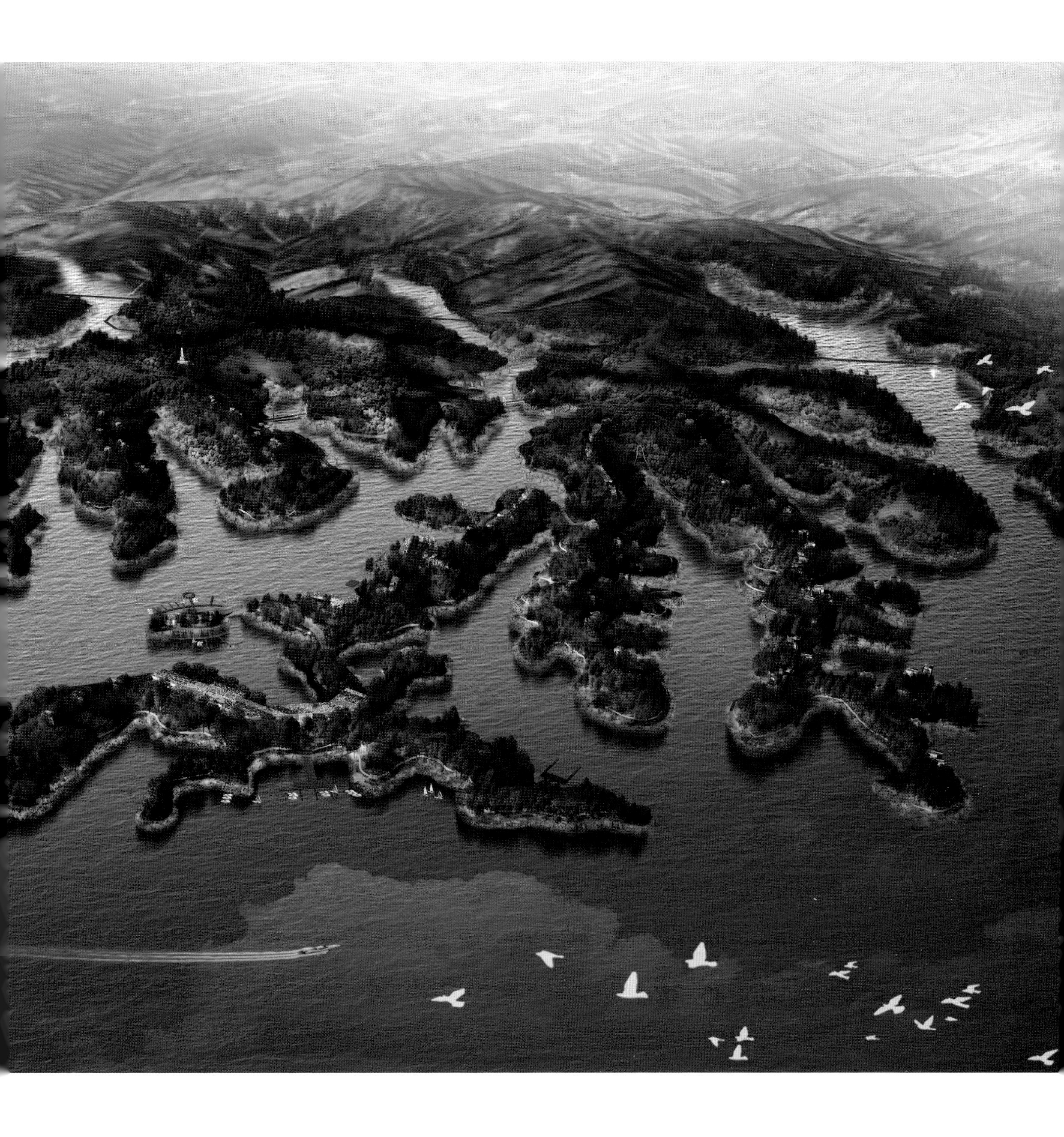

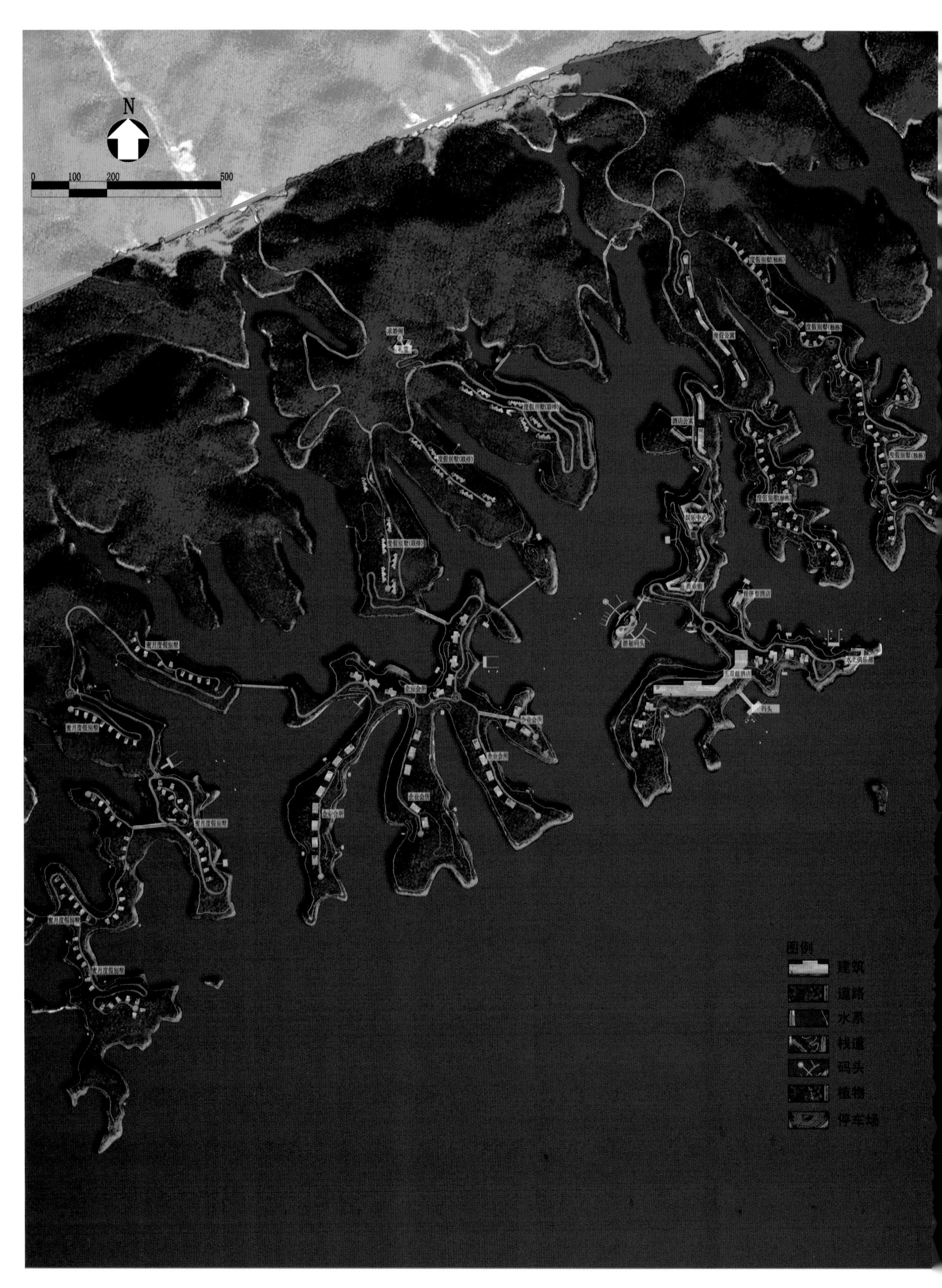
N
0
100
200
500
求婚阁
礼堂
度假别墅(联排)
度假别墅(联排)
度假别墅(联排)
度假别墅(独栋)
度假公寓
度假别墅(独栋)
酒店公寓
度假别墅(独栋)
度假别墅(独栋)
娱乐中心
商业街
经济型酒店
游艇码头
水上俱乐部
五星级酒店
码头
蜜月度假别墅
企业会所
企业会所
蜜月度假别墅
企业会所
企业会所
企业会所
蜜月度假别墅
蜜月度假别墅
蜜月度假别墅
图例
建筑
道路
水系
栈道
码头
植物
停车场

美国明尼阿波利斯密西西比河滨水设计

项目地点：美国 明尼苏达州 明尼阿波利斯市

项目规模：450公顷

项目类别：城市设计、景观设计

委 托 方：明尼阿波利斯公园与休憩委员会

Mississippi Riverfront Design in Minneapolis, USA

Location: Minneapolis, Minnesota, USA

Area (size): $450hm^2$

Project Type: Urban Planning, Landscape Architecture

Client: Minneapolis Park and Recreation Board

秦皇岛海岸线重点区域景观设计及建筑设计

项目地点： 河北省 秦皇岛市

项目规模： 64公顷（建筑：1949平方米）

项目类别： 景观设计、建筑设计

委 托 方： 秦皇岛市园林局

Landscape Architecture and Architecture Design of Coastal Key Areas in Qinhuangdao

Location: Qinhuangdao, Hebei Province

Area (size): $64hm^2$ (architecture: $1949m^2$)

Project Type: Landscape Architecture, Architecture Design

Client: Gardening Bureau of Qinhuangdao

秦皇岛市红旗北路西侧绿地公园

项目地点：河北省 秦皇岛市
项目规模：22.7公顷
项目类别：详细规划
委 托 方：秦皇岛市规划局海港分局

Greenspace Park at West Side of Hongqi North Road in Qinhuangdao

Location: Qinhuangdao City, Hebei Province
Area (size): 22.7hm^2
Project Type: Detailed Planning
Client: Haigang Branch of Planning Bureau of Qinhuangdao

鹤壁市淇河生态公园景观工程

项目地点：河南省 鹤壁市

项目规模：2.1平方千米

项目类别：城市设计、景观设计、建筑设计

委 托 方：鹤壁市规划局

Landscape Engineering of Qihe Eco Park in Hebi

Location: Hebi City, Henan Province

Area (size): 2.1km^2

Project Type: Urban Design, Landscape Architecture, Architecture Design

Client: Planning Bureau of Hebi

连云港东河新城城市设计

项目地点：江苏省 连云港市
项目规模：360公顷
项目类别：城市设计
委 托 方：连云港建设局

Urban Design of Donghe New Town in Lianyungang

Location: Lianyungang City, Jiangsu Province
Area (size): $360hm^2$
Project Type: Urban Design
Client: Construction Bureau of Lianyungang

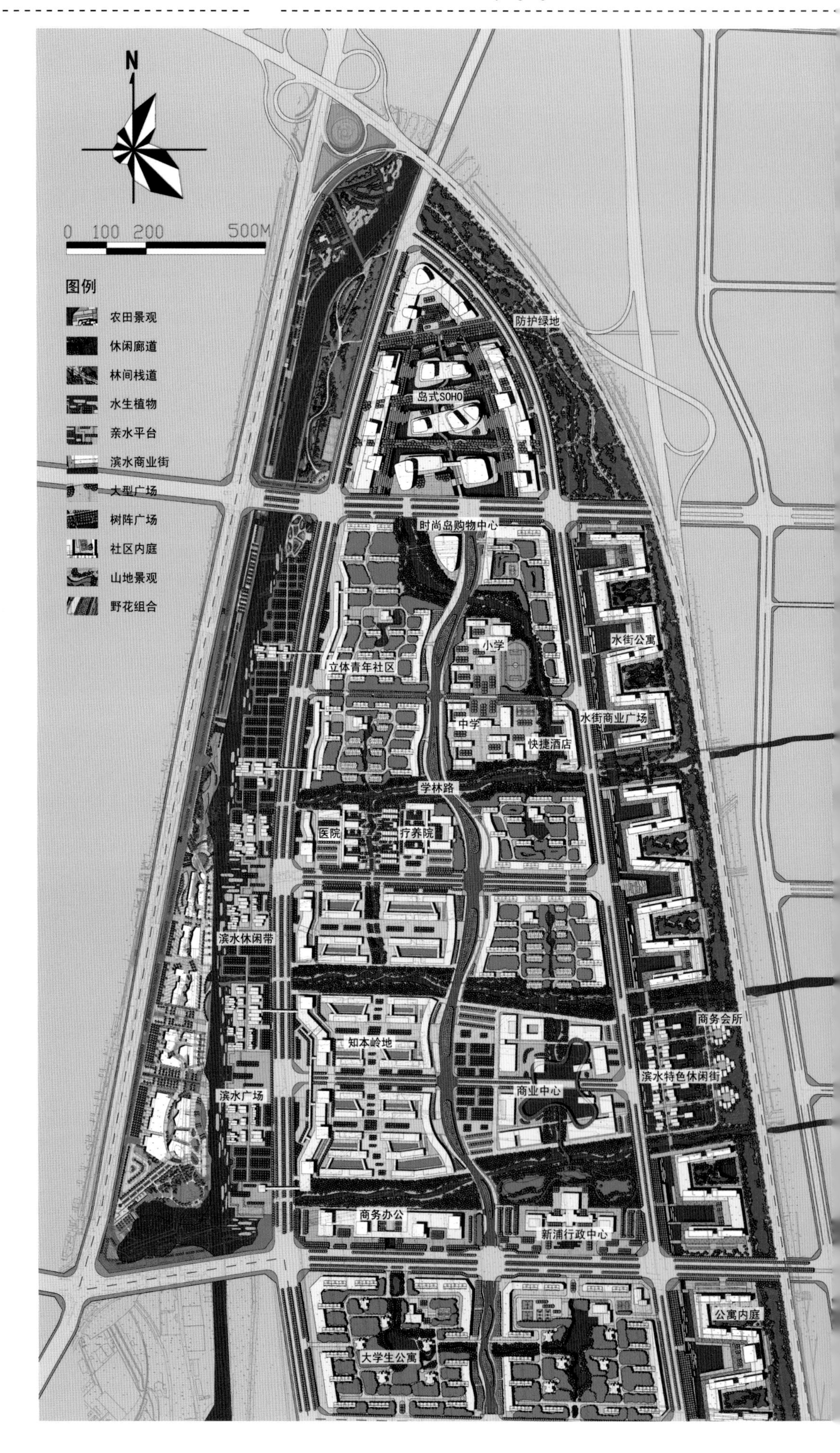

大连小窑湾景观设计

项目地点：辽宁省 大连市

项目规模：153.6公顷

项目类别：景观设计

委 托 方：大连市经济技术开发区重点工程建设办公室

Landscape Architecture of Xiaoyaowan, Dalian

Location: Dalian City, Liaoning Province

Area (size): 153.6hm^2

Project Type: Landscape Architecture

Client: Construction Office of Priority Projects in Dalian Economic and Technical Development District

烟台市区东北部岸线景观修建性详细规划

项目地点：山东省 烟台市
项目规模：6平方千米
项目类别：景观设计
委 托 方：烟台市市规划局

Constructive Detailed Planning of Northeastern Coastline Landscape of Yantai Downtown

Location: Yantai City, Shandong Province
Area (size): 6km^2
Project Type: Landscape Architecture
Client: Planning Bureau of Yantai

大同市十里河生态廊道

项目地点：山西省 大同市
项目规模：29千米河道两侧20-200米范围内
项目类别：城市规划、景观设计
委 托 方：山西省大同市规划管理局

Shilihe Eco Corridor in Datong

Location: Datong City, Shanxi Province
Area (size): 29km in length, 20~200 in width
Project Type: Urban Planning, Landscape Architecture
Client: Planning Bureau of Datong, Shanxi Province

上海后滩公园

项目地点：上海市 浦东区
项目规模：15公顷
项目类别：景观设计
委 托 方：上海世博局

Houtan Park, Shanghai

Location: Pudong, Shanghai
Area (size): 15hm^2
Project Type: Landscape architect
Client: Bureau of Shanghai Expo

N
0m 25m 50m 100m
水门码头广场
水门码头
后滩雨水泵站
污水泵站
原有建筑(隧道通风井)
图例:
原生湿地
保留建筑
亲水平台
植物泡泡
滨江密林
亲水台阶
广场铺装
广场树阵
田际折线步道
水生植条
梯地禾田
滨江休闲汀步
机器记忆
便捷步道
红绸带座椅
保留建筑构筑物
停车场
机器的容器
驳石护岸
田间小路

天津北塘地区北区景观设计

项目地点：天津市 滨海新区

项目规模：约400公顷

项目类别：景观设计

Landscape Architecture of North Tanggu District, Tianjin

Location: Binhai New Town, Tianjin

Area (size): Approximately 400hm^2

Project Type: Landscape Architecture

天津海河教育园区（北洋园）一期

项目地点：天津市 津南区

项目规模：228.39公顷

项目类别：景观设计

委 托 方：天津北洋园投资开发有限公司

Phase 1 of Haihe Education Area (Beijing Garden) in Tianjin

Location: Jinnan, Tianjin

Area (size): 228.39hm^2

Project Type: Landscape Architecture

Client: Tianjin Beiyang Garden Investment Development Co. Ltd.

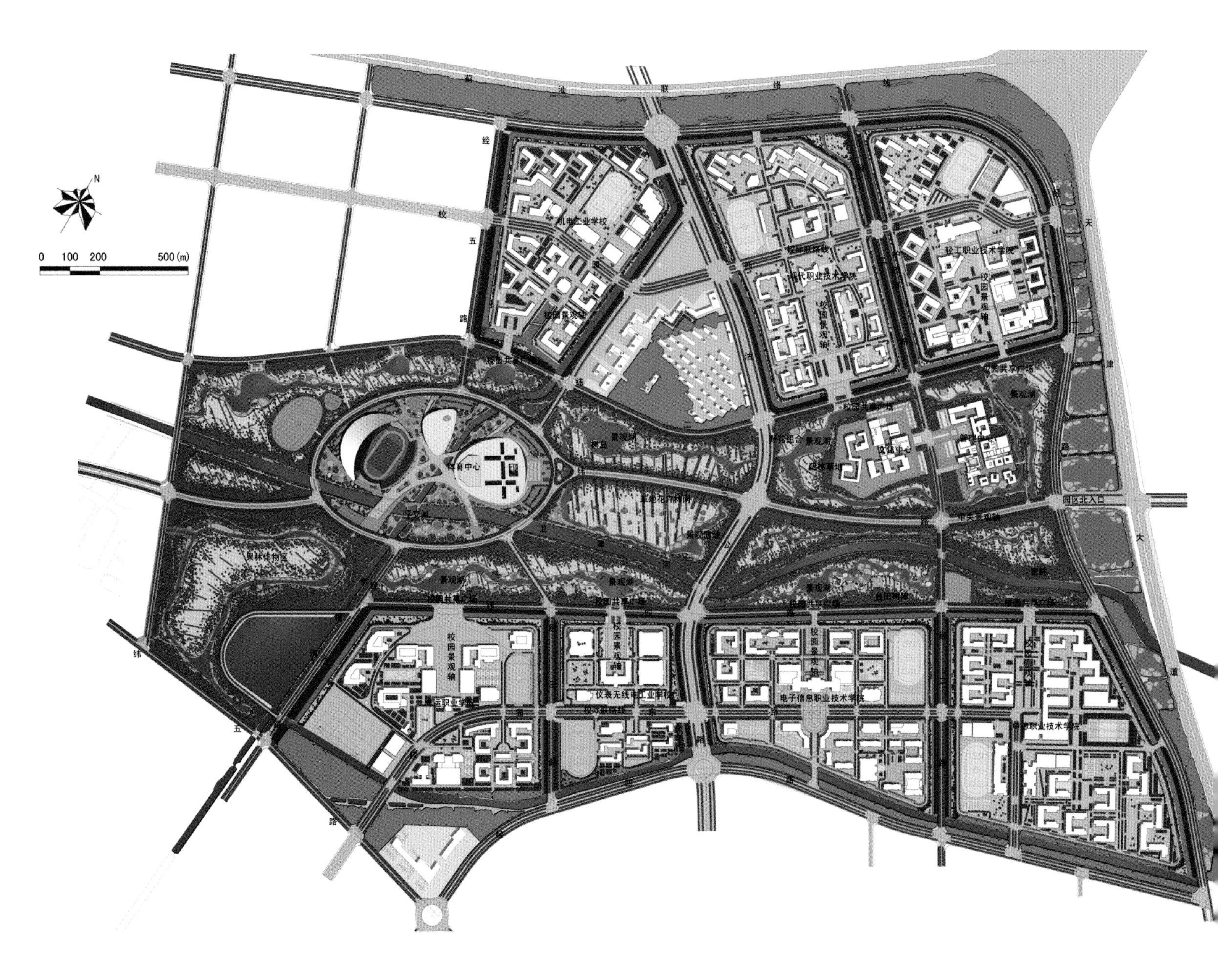

天津市东丽湖滨水景观规划及会议半岛详细城市设计

项目地点：天津市 东丽区
项目规模：408公顷
项目类别：城市设计
委 托 方：天津市东丽湖旅游开发总公司

Landscape Architecture of Dongli Lake Waterfront and Detailed Urban Design of Huiyi Peninsula

Location: Dongli, Tianjin
Area (size): 408hm²
Project Type: Urban Design
Client: Tianjin Dongli Lake Tourist Development Co. Ltd.

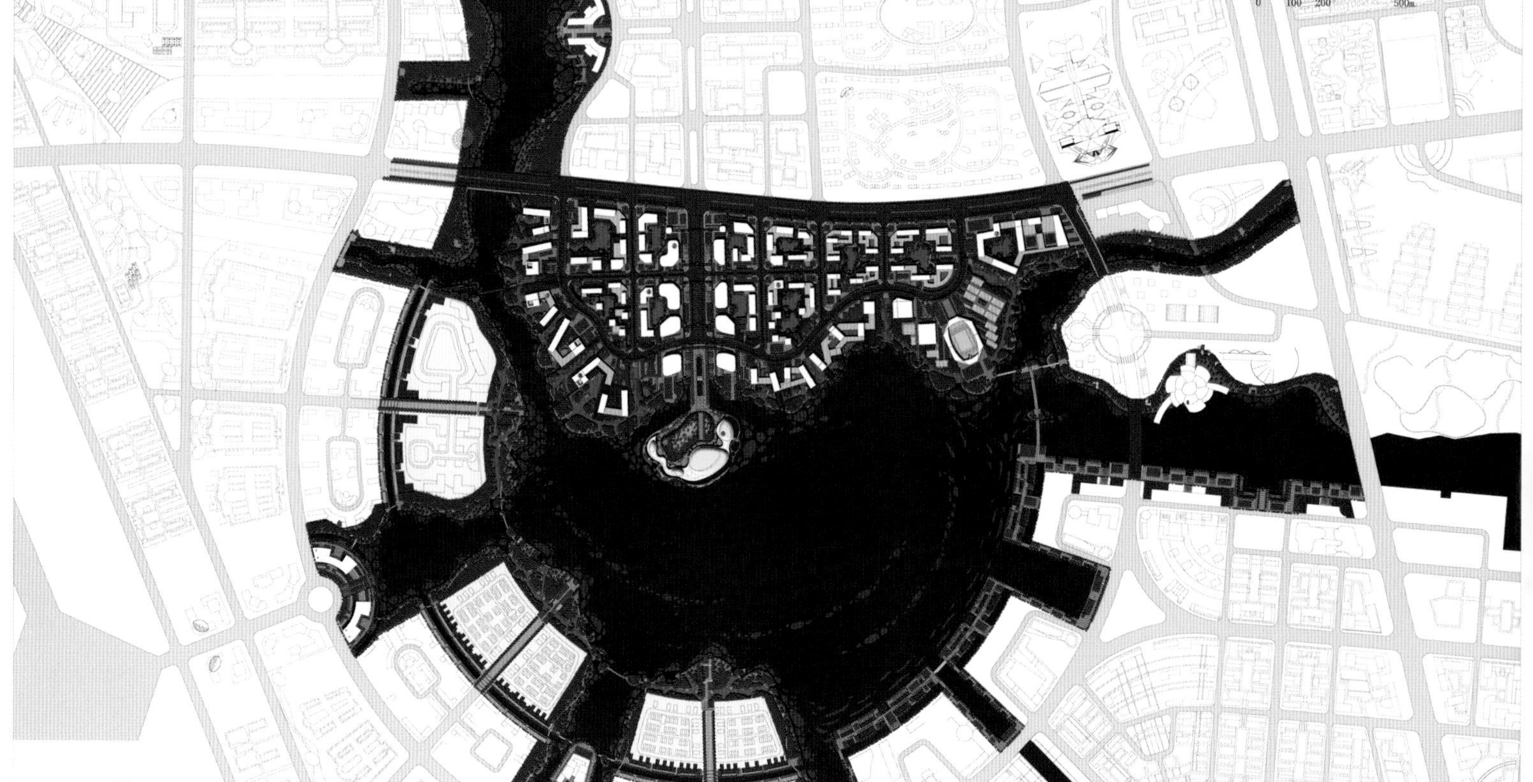

浙江省宁波梅山保税港区中央运动公园城市设计

项目地点：宁波市 梅山岛

项目规模：66公顷

项目类别：城市规划、景观设计

委 托 方：宁波梅山保税港区规划局

Urban Design of Central Activity Park of Meishan Bonded Port in Ningbo, Zhejiang

Location: Meishan Island, Ningbo City

Area (size): 66hm^2

Project Type: Urban Planning, Landscape Architecture

Client: Planning Bureau of Ningbo Meishan Bonded Port

浙江三门滨海新城横港及金鳞湖景观设计

项目地点：浙江省 三门县

项目规模：400公顷

项目类别：城市设计、景观设计、建筑设计

委 托 方：三门县滨海新城管理委员会

Landscape Architecture of Coastal New Town Heng Port and Jilin Lake in Sanmen, Zhejiang Province

Location: Sanmen, Zhejiang Province

Area (size): 400hm^2

Project Type: Urban Design, Landscape Architecture, Architecture deisgn

Client: Coastal New Town Council of Sanmen County

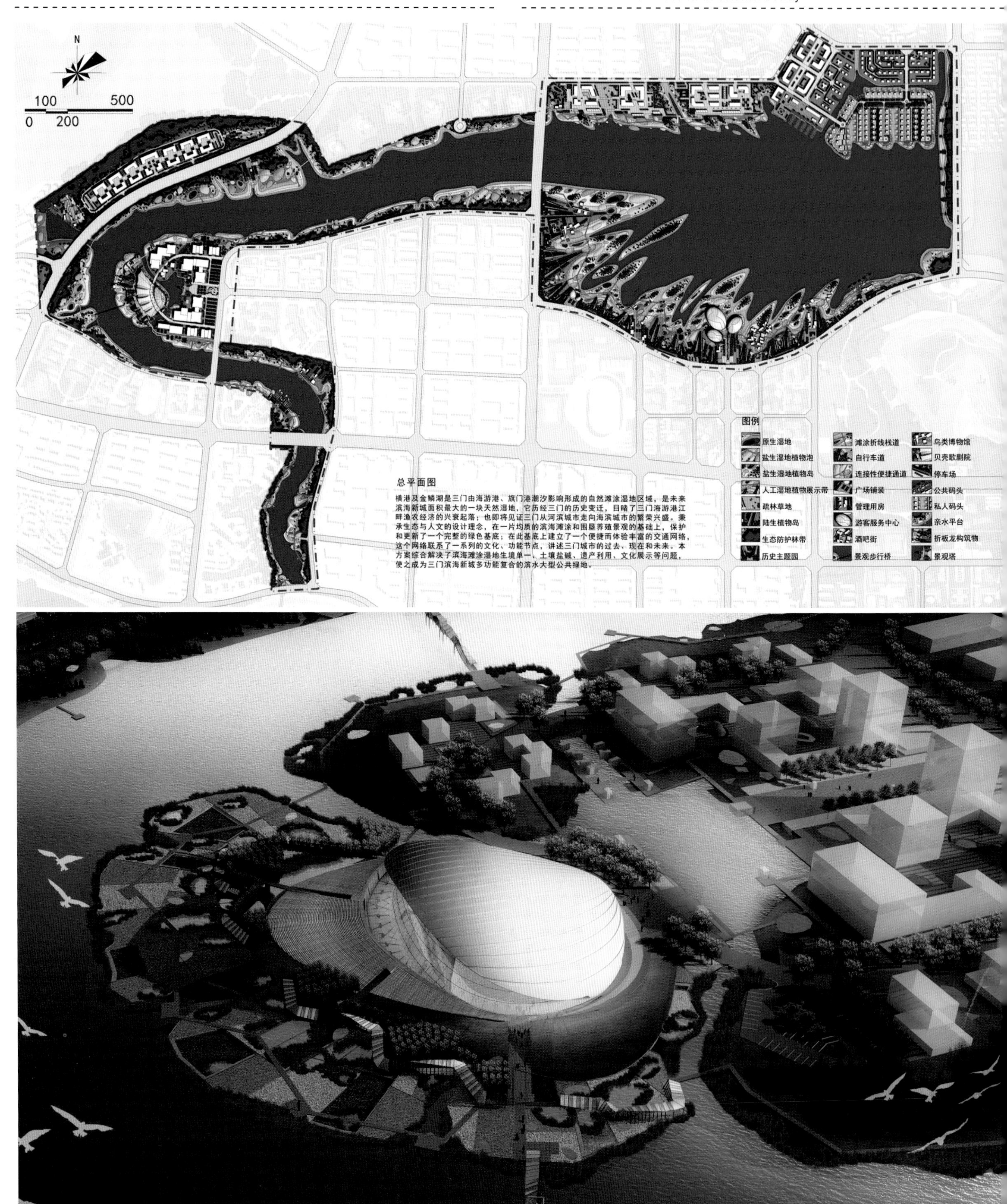

总平面图

横港及金鳞湖是三门由海游港、旗门港潮汐影响形成的自然滩涂湿地区域，是未来滨海新城面积最大的一块天然湿地，它历经三门的历史变迁，目睹了三门海游港江畔渔农经济的兴衰起落；也即将见证三门从河滨城市走向海滨城市的繁荣兴盛。秉承生态与人文的设计理念，在一片均质的滨海滩涂和围垦养殖景观的基础上，保护和更新了一个完整的绿色基底；在此基底上建立了一个便捷而体验丰富的交通网络，这个网络联系了一系列的文化、功能节点，讲述三门城市的过去、现在和未来。本方案综合解决了滨海滩涂湿地生境单一、土壤盐碱、遗产利用、文化展示等问题，使之成为三门滨海新城多功能复合的滨水大型公共绿地。

芝加哥北格兰特公园

项目地点：美国 芝加哥市
项目规模：10.7公顷
项目类别：景观设计
委 托 方：芝加哥市公园管理局

North Grant Park in Chicago

Location: Chicago, USA
Area (size): 10.7hm^2
Project Type: Landscape Architecture
Client: Chicago Park Conservancy

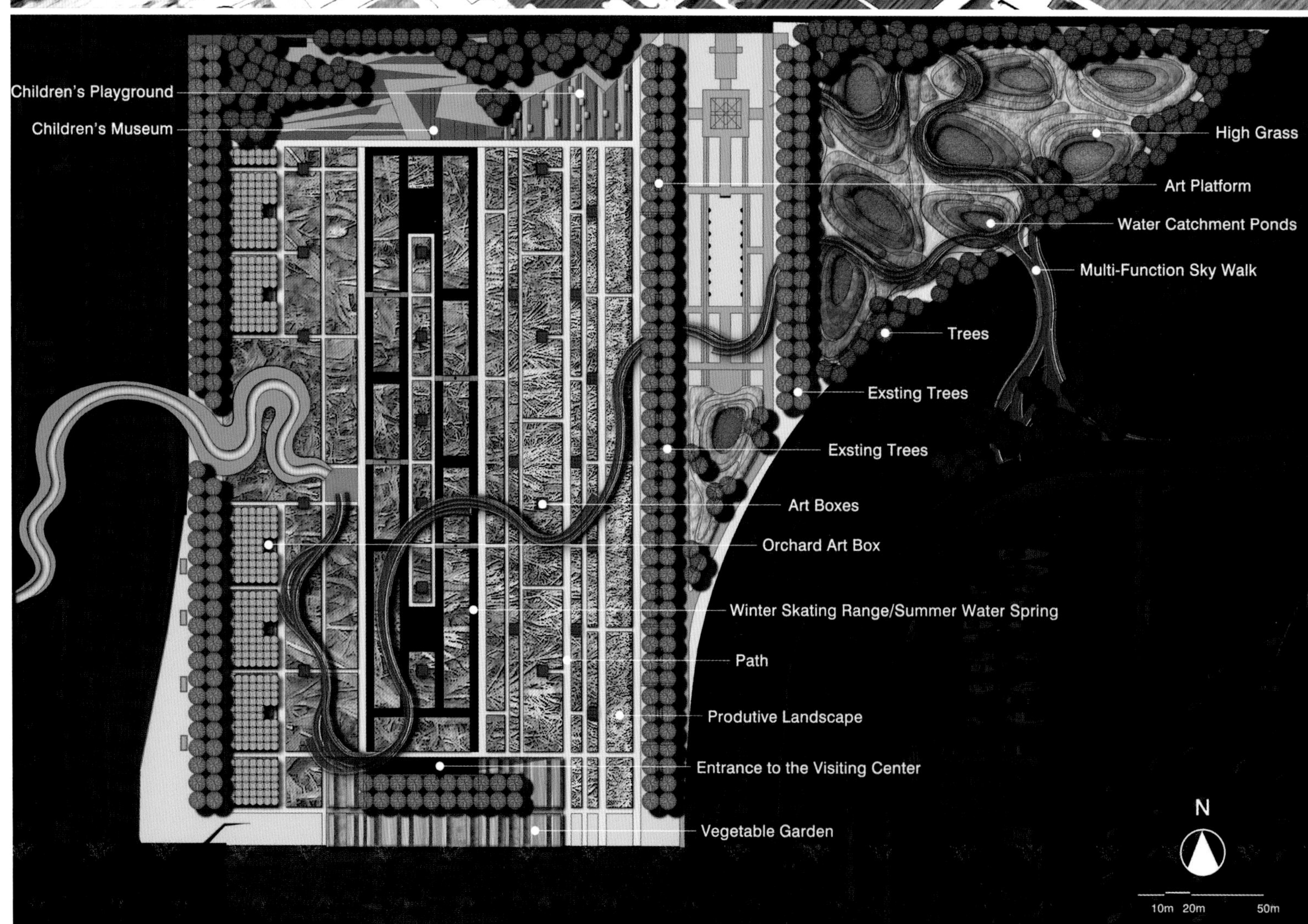

宜昌市运河公园

项目地点：湖北省 宜昌市
项目规模：11.92公顷
项目类别：景观设计、建筑设计
委 托 方：宜昌市城市投资建设开发有限公司

Channel Park in Yichang

Location: Yichang, Hubei Province
Area (size): 11.92hm^2
Project Type: Landscape Architecture, Architecture Design
Client: Yichang Urban Investment and Construction Development Co. Ltd.

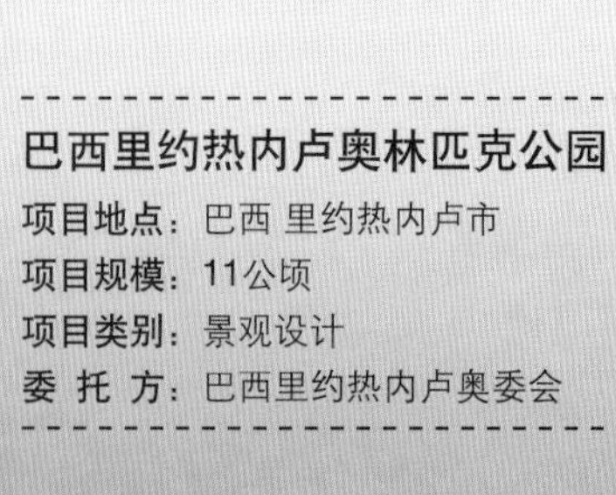

巴西里约热内卢奥林匹克公园

项目地点：巴西 里约热内卢市
项目规模：11公顷
项目类别：景观设计
委 托 方：巴西里约热内卢奥委会

Olympic Park in Rio, Brazil

Location: Rio, Brazil
Area (size): 11hm^2
Project Type: Landscape Architecture
Client: Brazil Rio Olympic Committee

吉林长春南关区光明公园景观工程

项目地点：吉林省 长春市

项目规模：39.5公顷

项目类别：景观设计

委 托 方：长春市南关区园林管理处

Landscape Engineering of Guangming Park in Nanguan District of Changchun, Jilin

Location: Changchun, Jilin Province

Area (size): $39.5hm^2$

Project Type: Landscape Architecture

Client: Garden Conservancy of Nanguan District of Changchun

山西大同晋华宫矿矿山公园概念规划

项目地点：山西省 大同市
项目规模：46.29公顷
项目类别：景观设计
委 托 方：大同市规划局

Concept Planning of Mine Park at Jinhuagong Mine in Datong, Shanxi

Location: Datong, Shanxi Province
Area (size): 46.29hm^2
Project Type: Landscape Architecture
Client: Planning Bureau of Datong

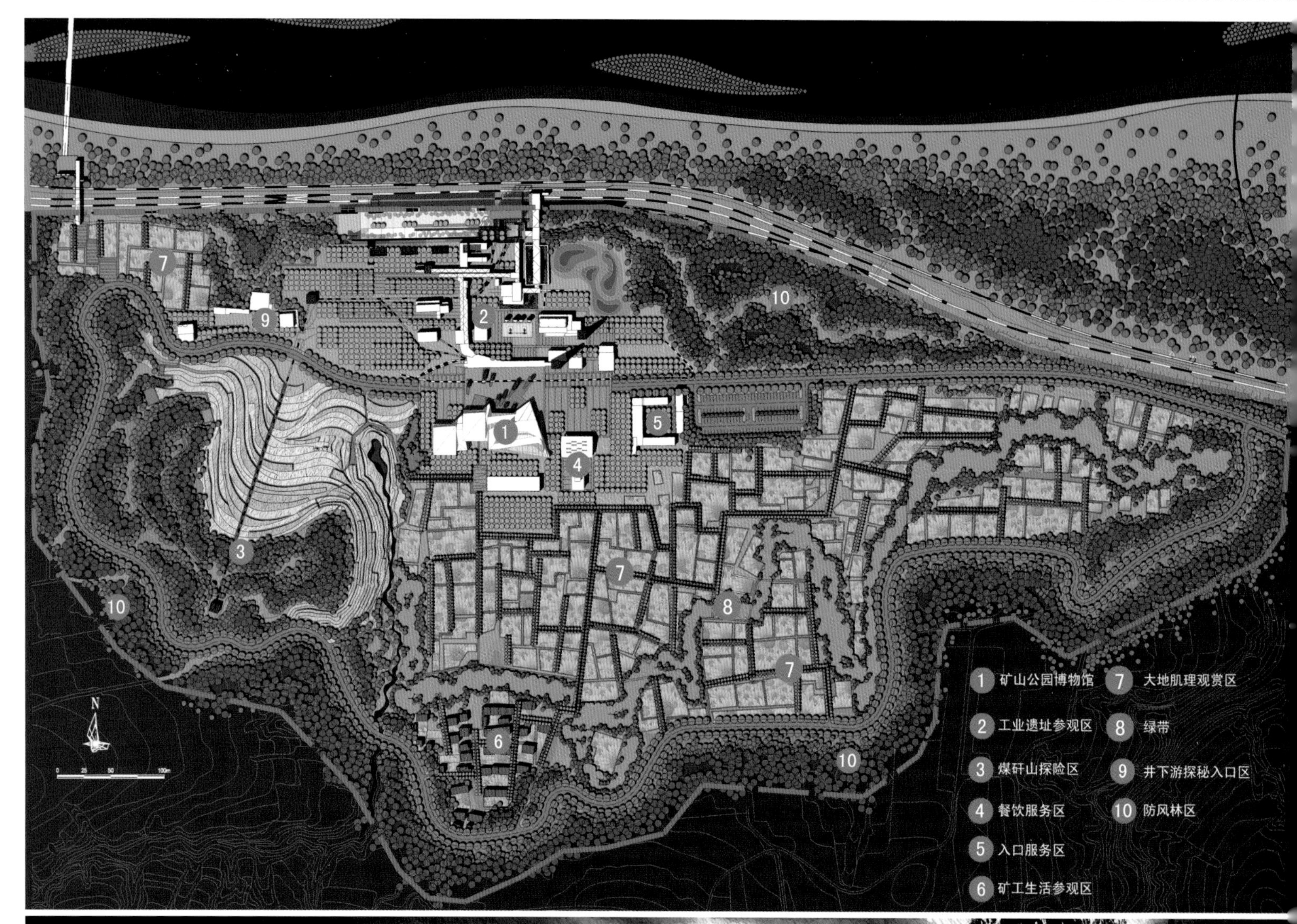

础 基 的 福 幸 户 万 家 千 是 全 安

晋
矿国家矿山公

西班牙马德里公园

项目地点：西班牙 马德里市

项目规模：11公顷

项目类别：景观设计

Madrid Park, Spain

Location: Madrid, Spain

Area (size): $11hm^2$

Project Type: Landscape Architecture

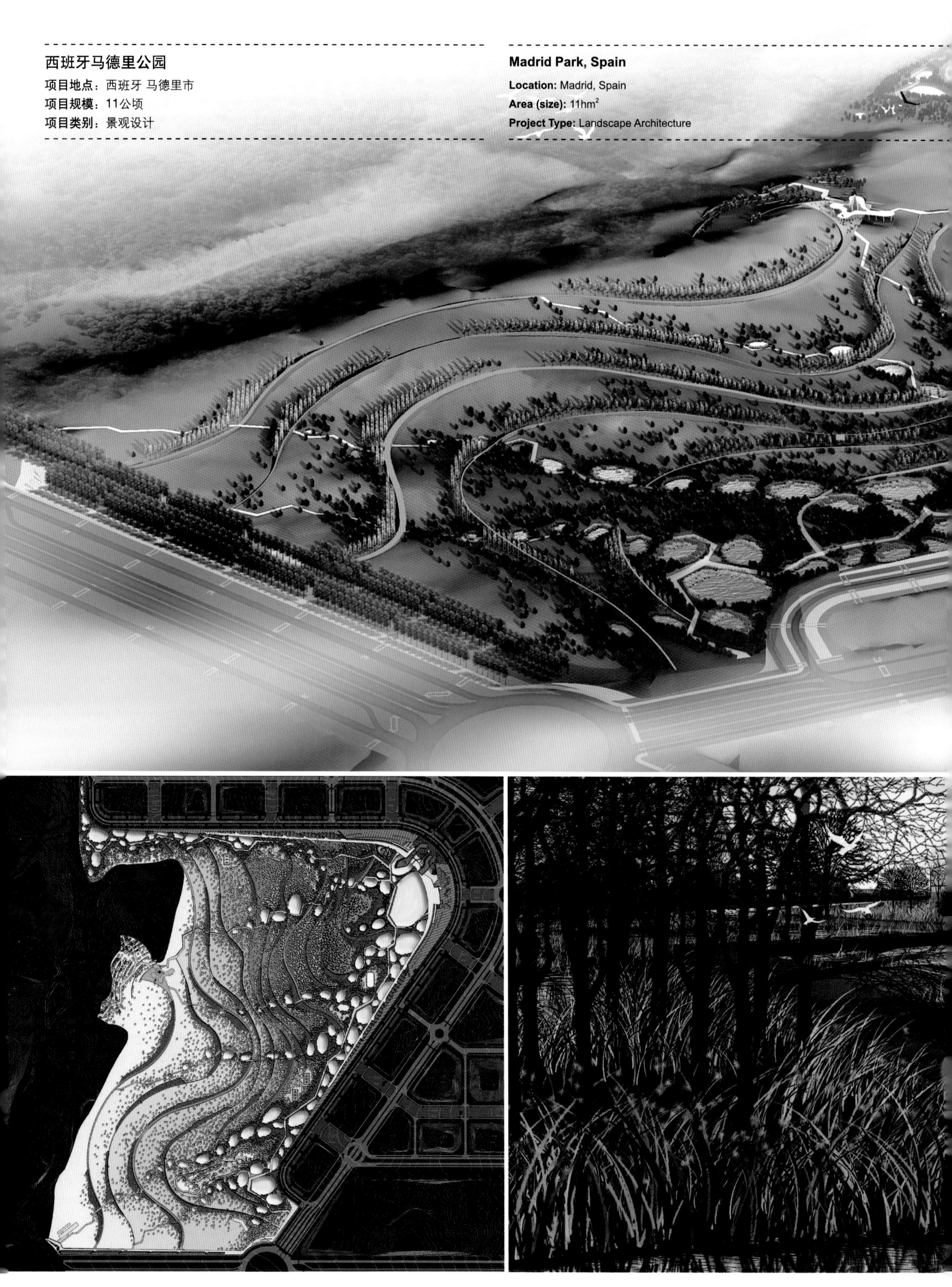

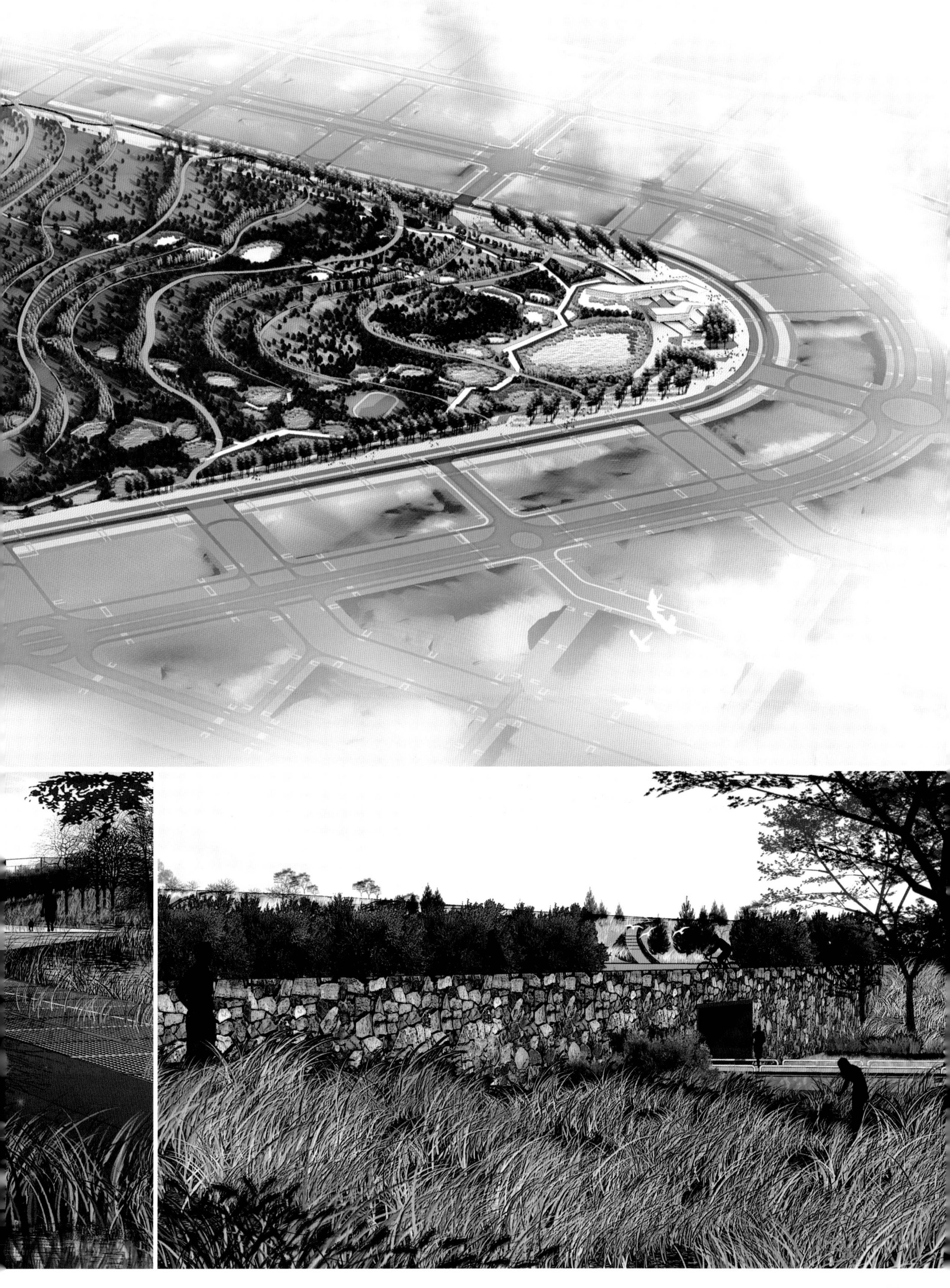

哈尔滨群力国家城市湿地公园

项目地点：哈尔滨市 群力区
项目规模：34公顷
项目类别：景观设计
委 托 方：哈尔滨群力新区开发建设管理办公

Qunli National Urban Wetland, Harbin

Location: [illegible] District, [illegible]
Area (size): 34hm^2
Project Type: Landscape Architecture
Client: Harbin Qunli Development and Construction Management Office

苏州高新区真山公园
项目地点：苏州市 高新区
项目规模：70公顷
项目类别：景观设计
委 托 方：苏州
Zhenshan Park in High-Tech District, Suzhou
Location: High-Tech District, Suzhou
Area (size): 70hm²
Project Type: Landscape Architecture
Client: Suzhou Tongan Tourist Development Co., Ltd.
0 50m 100m 150m 200m 250m
生态湿地
山地
园区道路
草地
岩石
湿地
水体
疏林广场
建筑
栈道
亲水平台
观赏廊架
台地密林
禾本科植物展示园
1 综合商业服务建筑
2 山体修复景观带
3 公园主入口
4 公园次入口
5 岩石地貌展示园
6 密林
7 生态湿地
8 观赏廊架
9 台地密林
10 现状山体
11 禾本科植物展示园
12 古墓展示节点
13 博物馆
14 架空廊道
15 观景挑台
16 水体景观
17 湖心岛
18 公园管理处
19 现状加油站
20 停车场
21 星级酒店
22 步行街
23 停车场
24 产权式度假酒店

郑州高新区彩虹花园景观改造工程

项目地点：河南省 郑州市

项目规模：6.88公顷

项目类别：景观设计

委 托 方：郑州高新区管委会

Landscape Transforming Project of Rainbow Garden in Zhengzhou High & New Technology Industries Development Zone

Location: Zhengzhou City, Henan Province

Area (size): 6.88hm^2

Project Type: Landscape Architecture

Client: Administrative Committee of Zhengzhou High & New Technology Industries Development Zone

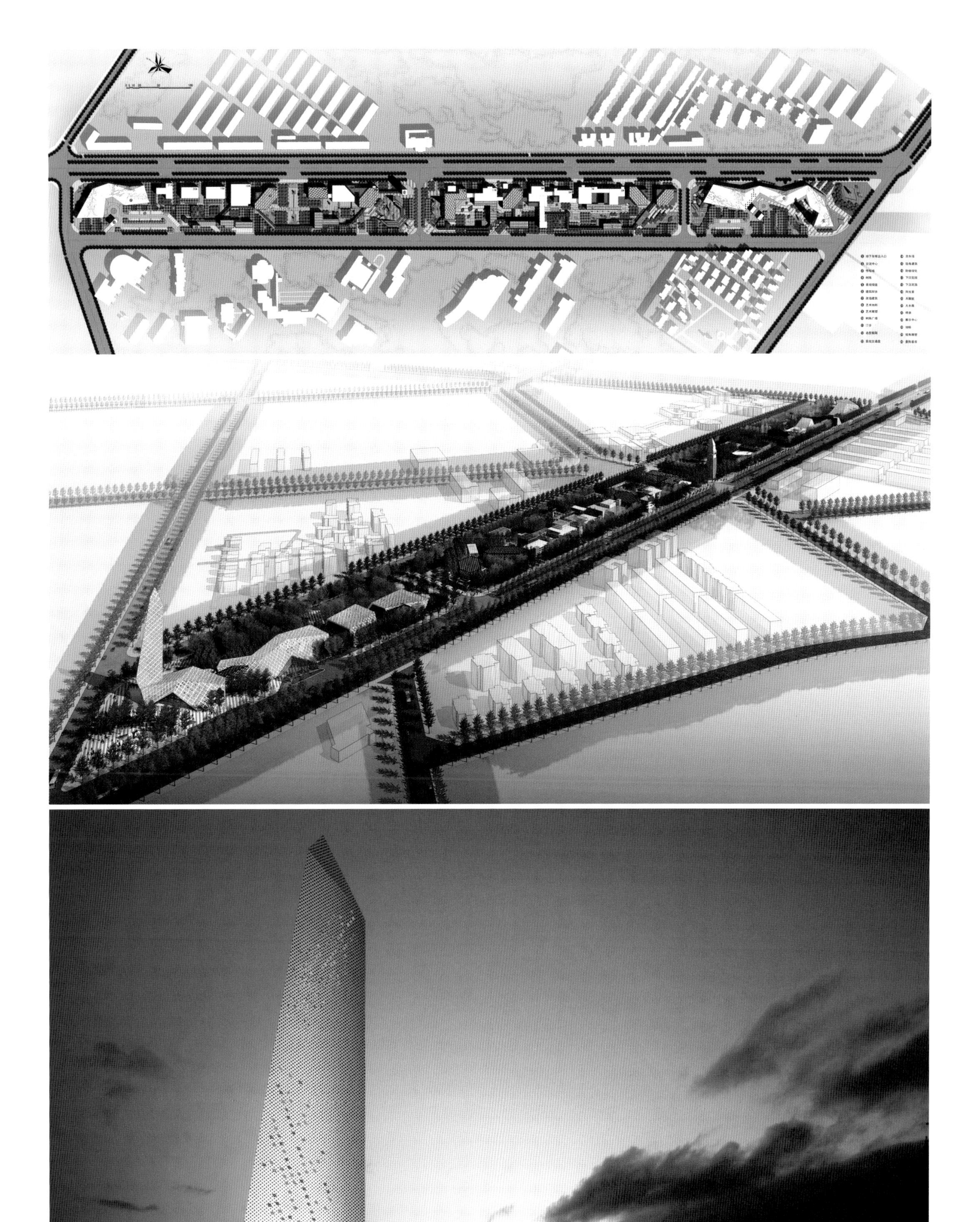

睢宁县新市街景观与商业建筑改造

项目地点：江苏省 睢宁县
项目规模：49万平方米
项目类别：城市设计、建筑设计、景观设计
委 托 方：睢宁县规划局

Landscape and Commercial Architecture Transforming of Xinshi Street in Suining

Location: Suining County, Jiangsu Province
Area (size): 490,000m^2
Project Type: Urban Design, Architecture Design, Landscape Architecture
Client: Suining County Planning Bureau

SHOPPING MALL
Dior
DIORIFIC
BOSS

大同煤气厂景观改造工程

项目地点：山西省 大同市

项目规模：26公顷

项目类别：景观设计

委 托 方：大同市规划局

Landscape Transforming Project of Datong Gas Plant

Location: Datong City, Shanxi Province

Area (size): 26hm^2

Project Type: Landscape Architecture

Client: Datong Planning Bureau

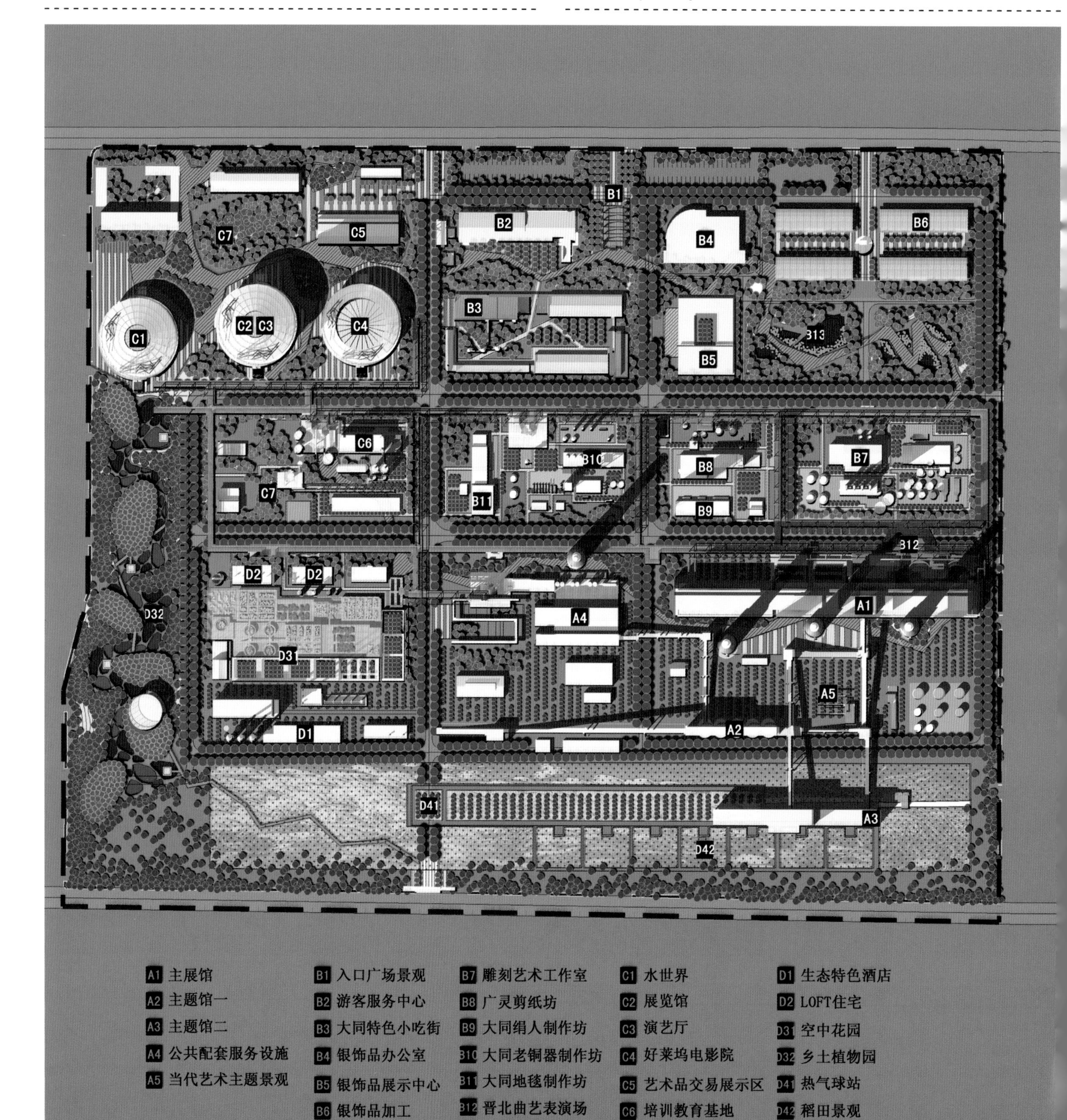

OSCAR

中国宣纸传习基地修建性详细规划

项目地点：安徽省 泾县
项目规模：27.3公顷
项目类别：规划设计、建筑设计、景观设计
委 托 方：中国宣纸集团公司

Constructive Detail Planning of Rice Paper Pass and Learn Base

Location: Jing County, Anhui Province
Area (size): 27.3hm^2
Project Type: Planning Design, Architecture Design, Landscape Architecture
Client: China Rice Paper Group Corporation

北京中关村科技园区昌平园研发基地城市设计方案

项目地点：北京市 昌平区

项目规模：137公顷

项目类别：城市设计

委 托 方：北京市昌平区政府

Urban Design Scheme of Research & Development Base in Changping Sub park, Zhongguancun Science Park

Location: Changping District, Beijing

Area (size): $137hm^2$

Project Type: Urban Design

Client: Changping District Government, Beijing

承德市宽城民族教育园区总体规划

项目地点：河北省 承德市
项目规模：28.9公顷
项目类别：城市设计、建筑设计
委 托 方：宽城教育局

General Planning of National Education Park in Town of Kuancheng, Chengde City

Location: Chengde City, Hebei Province
Area (size): 28.9hm^2
Project Type: Urban Design, Architecture Design
Client: Education Bureau of Town of Kuancheng

秦皇岛市党校建筑初步设计、景观工程设计

项目地点：河北省 秦皇岛市

项目规模：6.8公顷（建筑：29518平方米）

项目类别：景观设计、建筑设计

委 托 方：秦皇岛市社会公益项目建设管理中心

Preliminary Design, Landscape Engineering Project of Party School in the City of Qinhuangdao

Location: Qinhuangdao City, Hebei Province

Area (size): 6.8hm^2 (architecture area: 29518m^2)

Project Type: Landscape Architecture, Architecture Design

Client: Social Welfare Projects Construction and Management Center in Qinhuangdao City

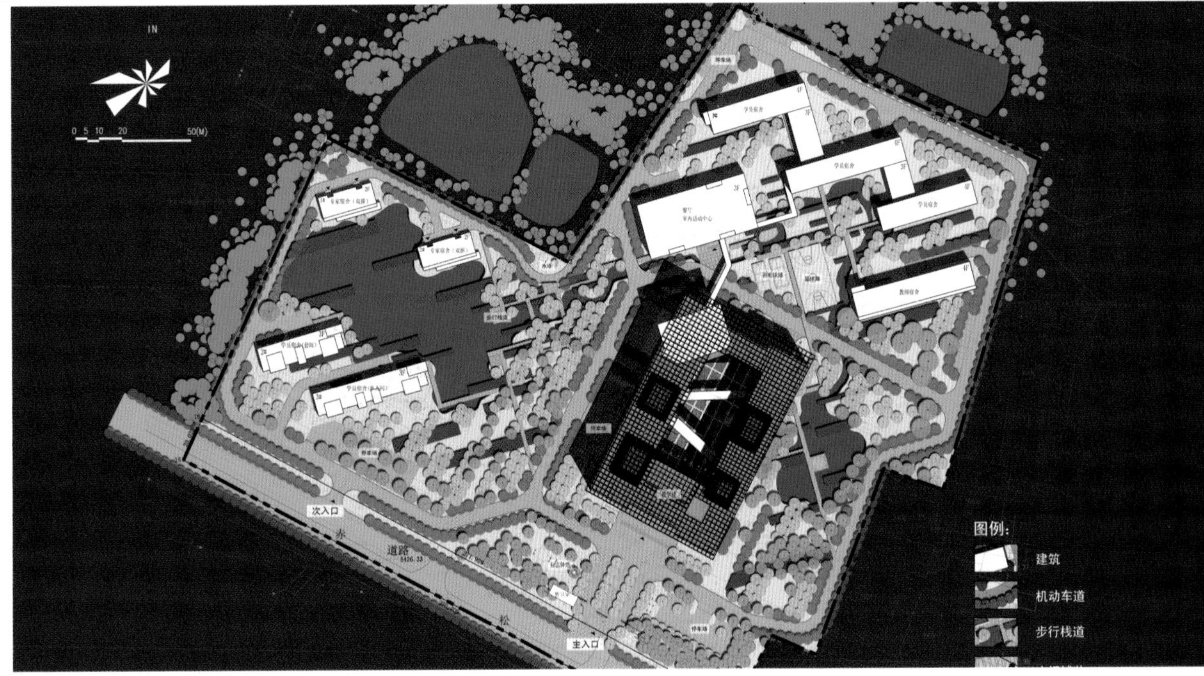
次入口
主入口
图例：
建筑
机动车道
步行栈道

洛阳师范学院校园总体规划方案设计

项目地点：河南省 洛阳市
项目规模：190公顷
项目类别：总体规划
委 托 方：洛阳师范大学

General Planning Scheme of the Campus of Luoyang Normal University

Location: Luoyang City, Henan Province
Area (size): 190hm^2
Project Type: General Planning
Client: Luoyang Normal University

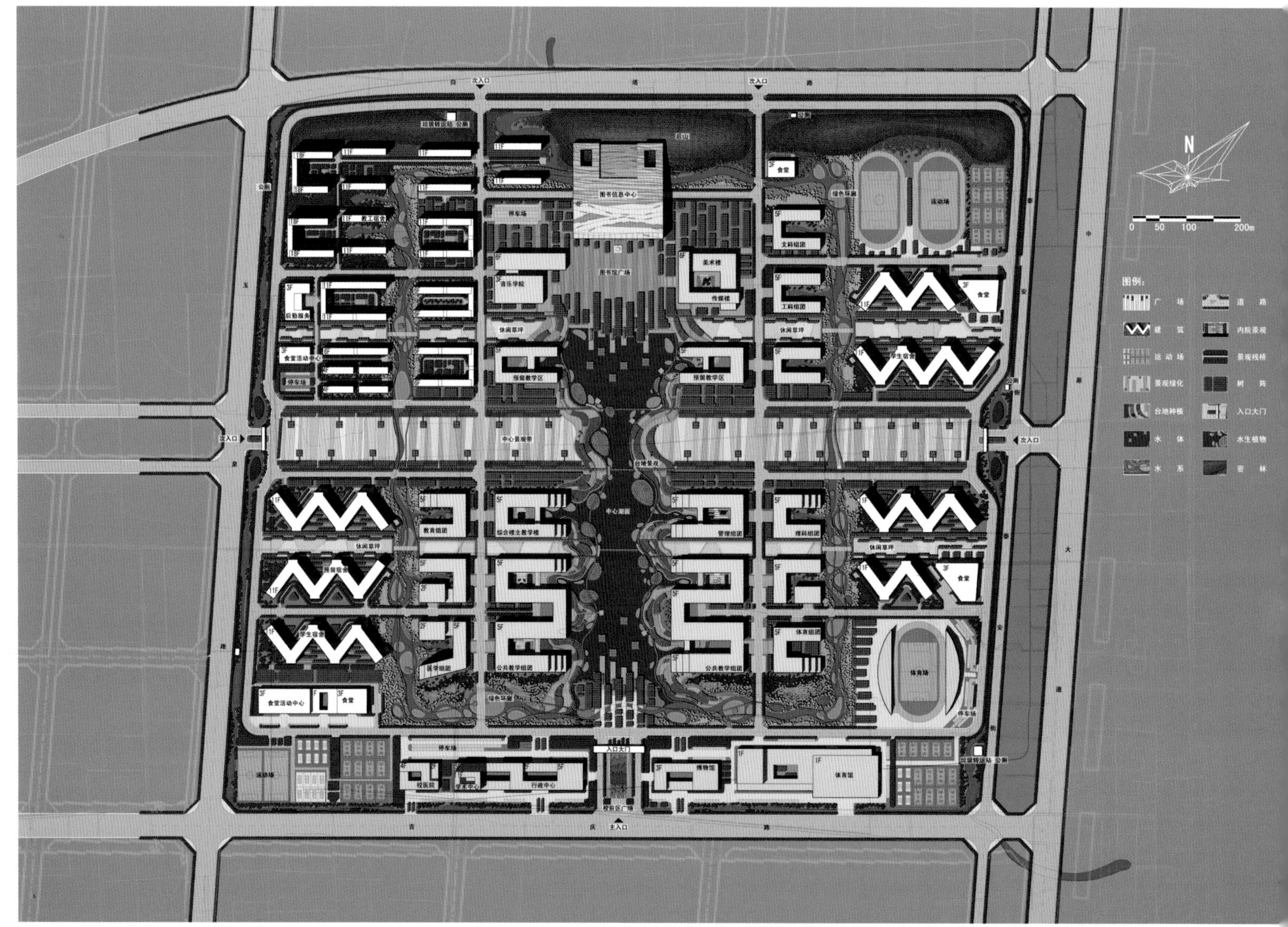

音乐系
The Department Of Mosic

美术系
The Department Of Art

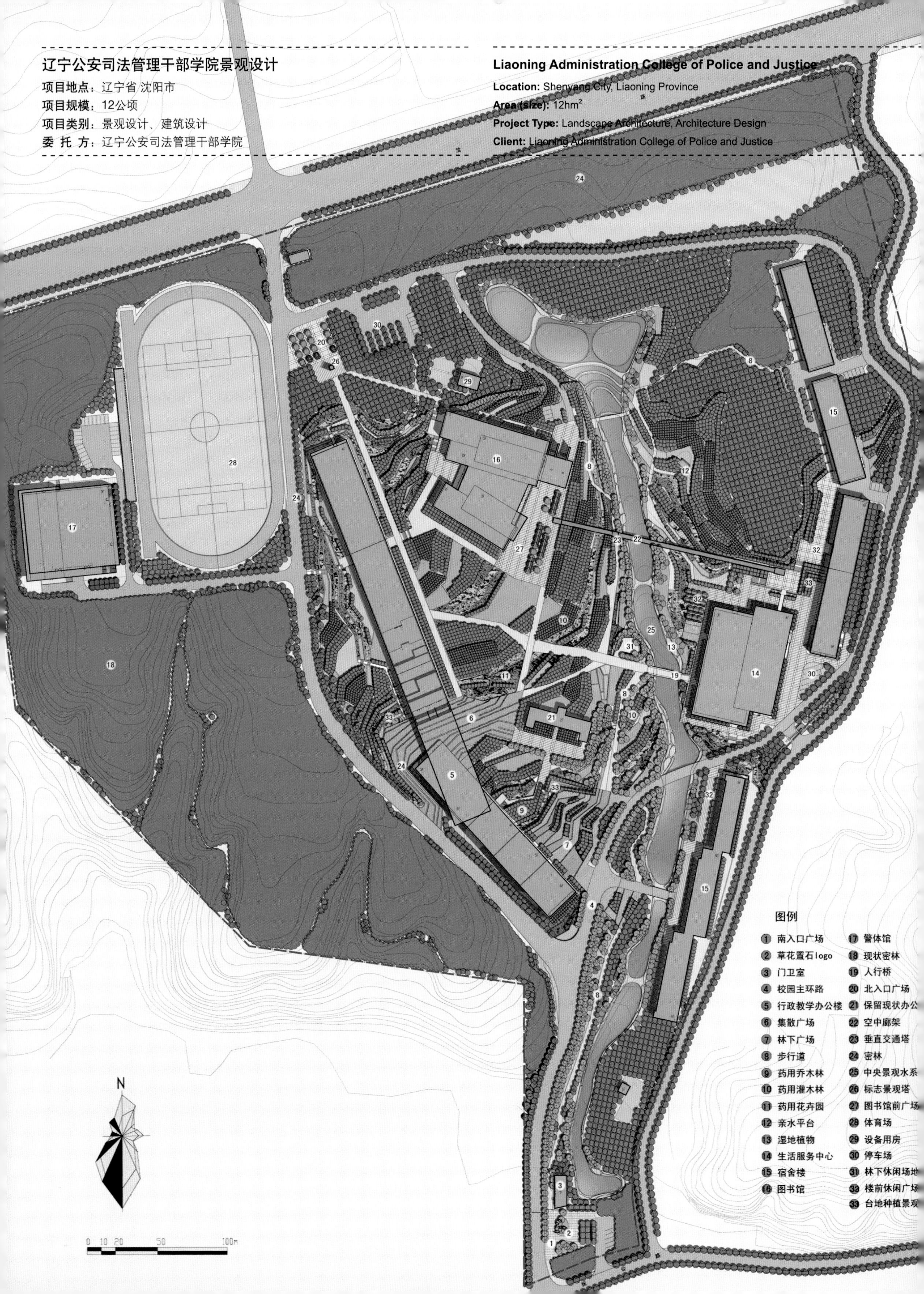
辽宁公安司法管理干部学院景观设计
项目地点：辽宁省 沈阳市
项目规模：12公顷
项目类别：景观设计、建筑设计
委 托 方：辽宁公安司法管理干部学院
Liaoning Administration College of Police and Justice
Location: Shenyang City, Liaoning Province
Area (size): 12hm²
Project Type: Landscape Architecture, Architecture Design
Client: Liaoning Administration College of Police and Justice
N
0 10 20 50 100m
图例
1 南入口广场
2 草花置石logo
3 门卫室
4 校园主环路
5 行政教学办公楼
6 集散广场
7 林下广场
8 步行道
9 药用乔木林
10 药用灌木林
11 药用花卉园
12 亲水平台
13 湿地植物
14 生活服务中心
15 宿舍楼
16 图书馆
17 警体馆
18 现状密林
19 人行桥
20 北入口广场
21 保留现状办公
22 空中廊架
23 垂直交通塔
24 密林
25 中央景观水系
26 标志景观塔
27 图书馆前广场
28 体育场
29 设备用房
30 停车场
31 林下休闲场地
32 楼前休闲广场
33 台地种植景观

云南红河技师学院概念规划及修建性详细规划

项目地点：云南省 红河哈尼族彝族自治州 开远市

项目规模：概念规划301.29公顷、修建性详细规划100.56公顷

项目类别：概念规划、修建性详细规划

委 托 方：红河技师学院领导小组办公室

Conceptual and Constructive Detail Planning of Honghe Technician Institute, Yunan Province

Location: Kaiyuan City, Honghe Hani & Yi Autonomous Prefecture, Yunan Province

Area (size): 301.29hm^2 (Conceptual Planning), 100.56hm^2 (Constructive Detail Planning)

Project Type: Conceptual Planning, Constructive Detail Planning

Client: Office of Leader Team, Honghe Technician Institute

N
0 40 80 160m
图例
规划建筑
现状保留建筑
农田
山体
密林
疏林草地
广场
湿生植物
水体
道路
规划范围
01 校园主入口
02 系楼
03 中央景观带
04 学生活动中心
05 公共教学楼
06 实验楼
07 研发鉴定中心
08 实训楼
09 梯田公园
10 大礼堂
11 图书馆
12 食堂
13 学生宿舍
14 生活服务设施
15 风雨操场
16 后勤保障中心
17 体育场
18 体育馆
19 主体育场
20 游泳馆
21 滨水休闲带
22 校园次入口
23 行政楼
24 景观农田

化工系

昆明理工大学呈贡校区规划设计

项目地点：云南省 昆明市
项目规模：133公顷
项目类别：规划、建筑设计
委 托 方：昆明理工大学

Planning Design of Chenggong Campus, Kunming University of Science and Technology

Location: Kunming City, Yunan Province
Area (size): 133hm^2
Project Type: Planning, Architecture Design
Client: Kunming University of Science and Technology

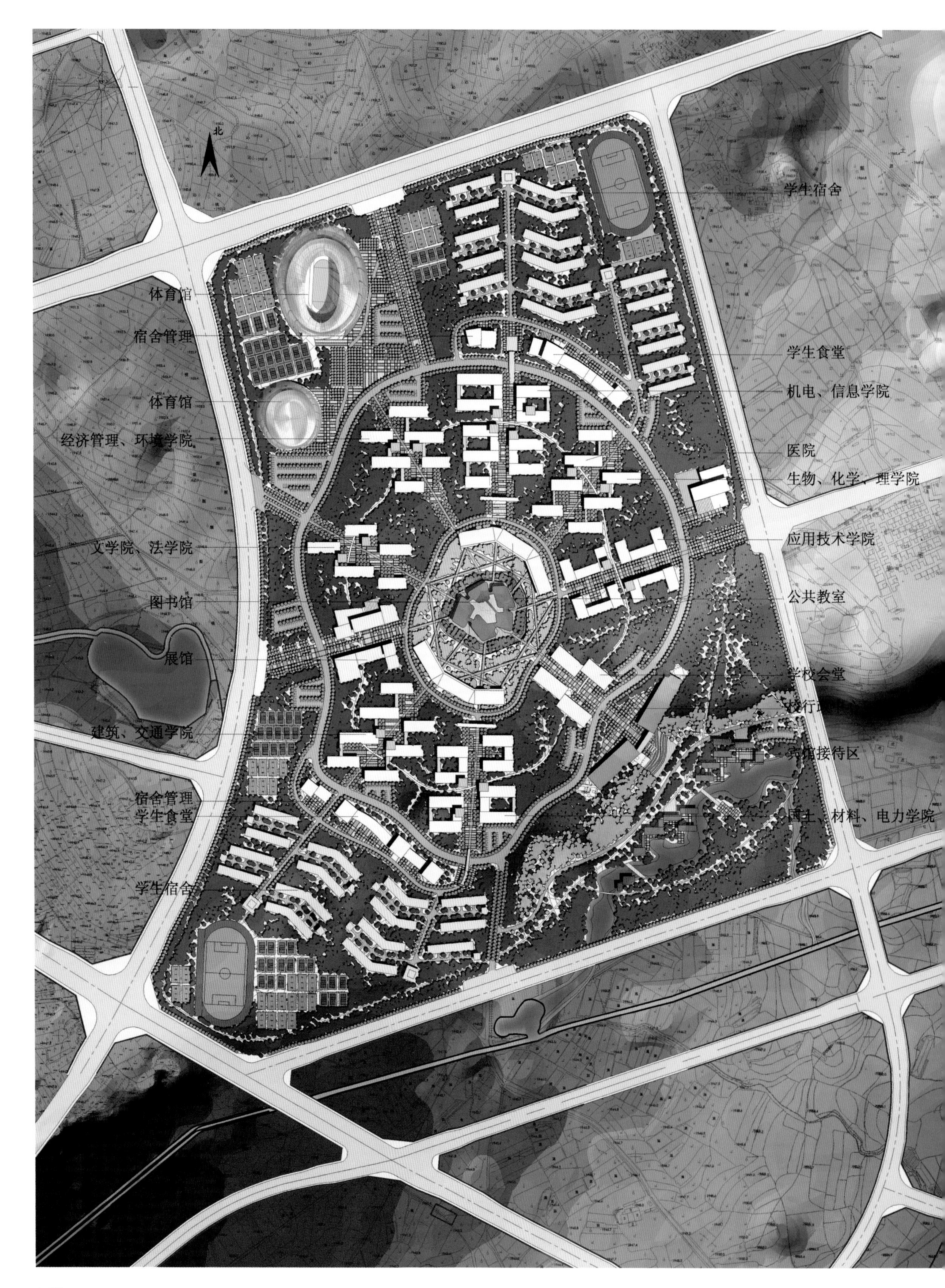
北
学生宿舍
体育馆
宿舍管理
学生食堂
机电、信息学院
体育馆
经济管理、环境学院
医院
生物、化学、理学院
文学院、法学院
应用技术学院
图书馆
公共教室
展馆
学校会堂
宾馆接待区
建筑、交通学院
宿舍管理
学生食堂
国土、材料、电力学院
学生宿舍

SOCIAL SCILENCE
NATURE SCILENCE
ARTS/SPORTS/ENTERTAMENT
CIAL SCIENCE
NATURE SECICENSE

宜春学院高安校区建筑工程设计

项目地点：江西省 宜春市 高安县

项目规模：46.67公顷

项目类别：建筑设计

委 托 方：宜春学院高安校区

Architectural Engineering Design of Gao'an Campus, the College of Yichu

Location: County of Gao'an, Yichun City, Jiangxi Province

Area (size): 46.67hm^2

Project Type: Architectural Design

Client: Gao'an Campus, the College of Yichun

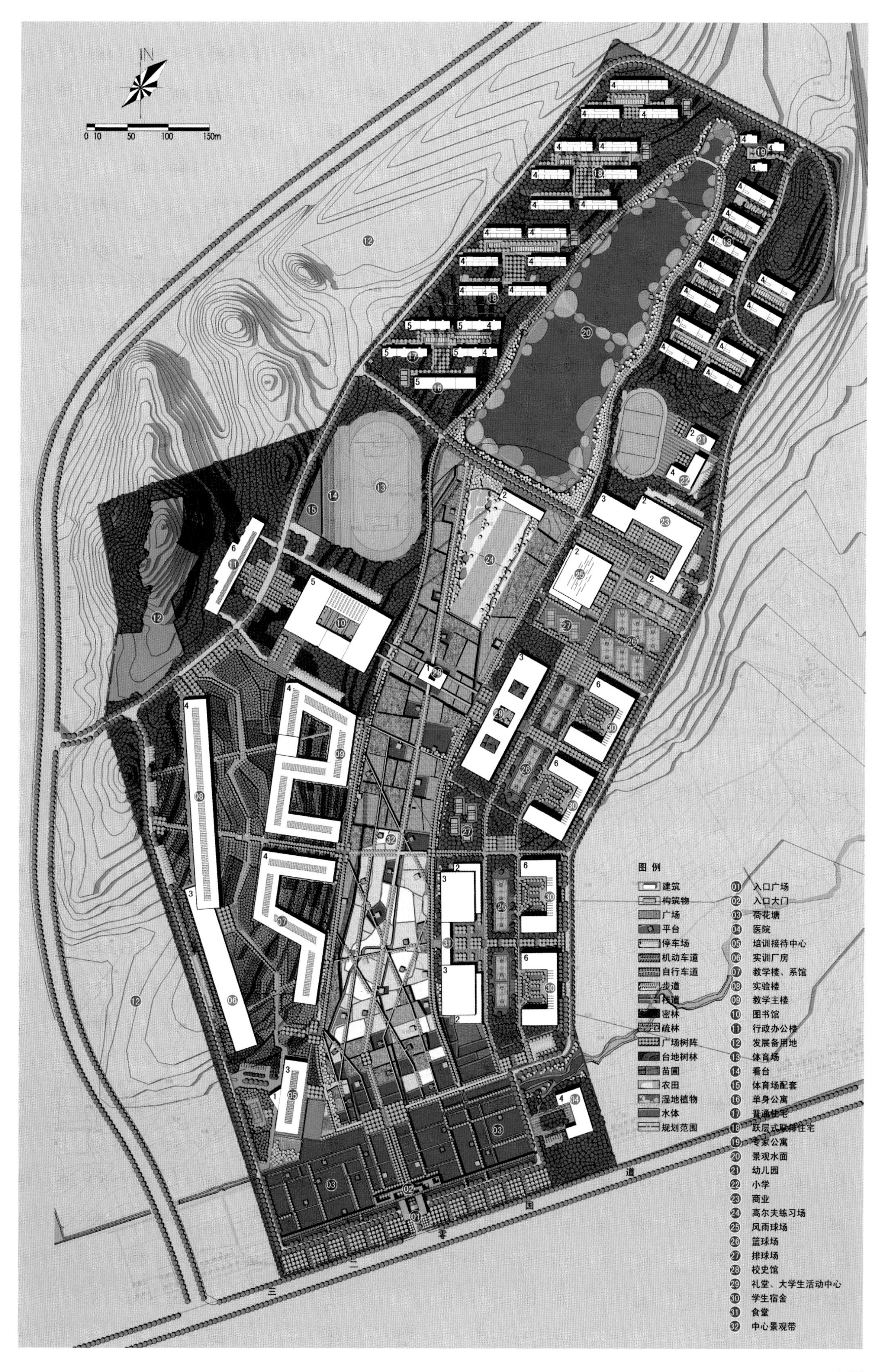
N
0 10 50 100 150m
图例
建筑
构筑物
广场
平台
停车场
机动车道
自行车道
步道
栈道
密林
疏林
广场树阵
台地树林
苗圃
农田
湿地植物
水体
规划范围
01 入口广场
02 入口大门
03 荷花塘
04 医院
05 培训接待中心
06 实训厂房
07 教学楼、系馆
08 实验楼
09 教学主楼
10 图书馆
11 行政办公楼
12 发展备用地
13 体育场
14 看台
15 体育场配套
16 单身公寓
17 普通住宅
18 跃层式联排住宅
19 专家公寓
20 景观水面
21 幼儿园
22 小学
23 商业
24 高尔夫练习场
25 风雨球场
26 篮球场
27 排球场
28 校史馆
29 礼堂、大学生活动中心
30 学生宿舍
31 食堂
32 中心景观带
三 二 零 国 道

林州红旗渠博物院建筑工程设计

项目地点：河南省 林州市

项目规模：博物苑8000平米，青年洞1000平米

项目类别：建筑设计

委 托 方：林虑山风景名胜区管委会

Architectural Engineering Design of Red Flag Canal Museum, Linzhou

Location: Linzhou City, Henan Province

Area (size): 8,000m^2 (Museum), 1,000m^2 (Youth Tunnel)

Project Type: Architectural Design

Client: Administrative Committee of Scenic Area of Linlv Mountain

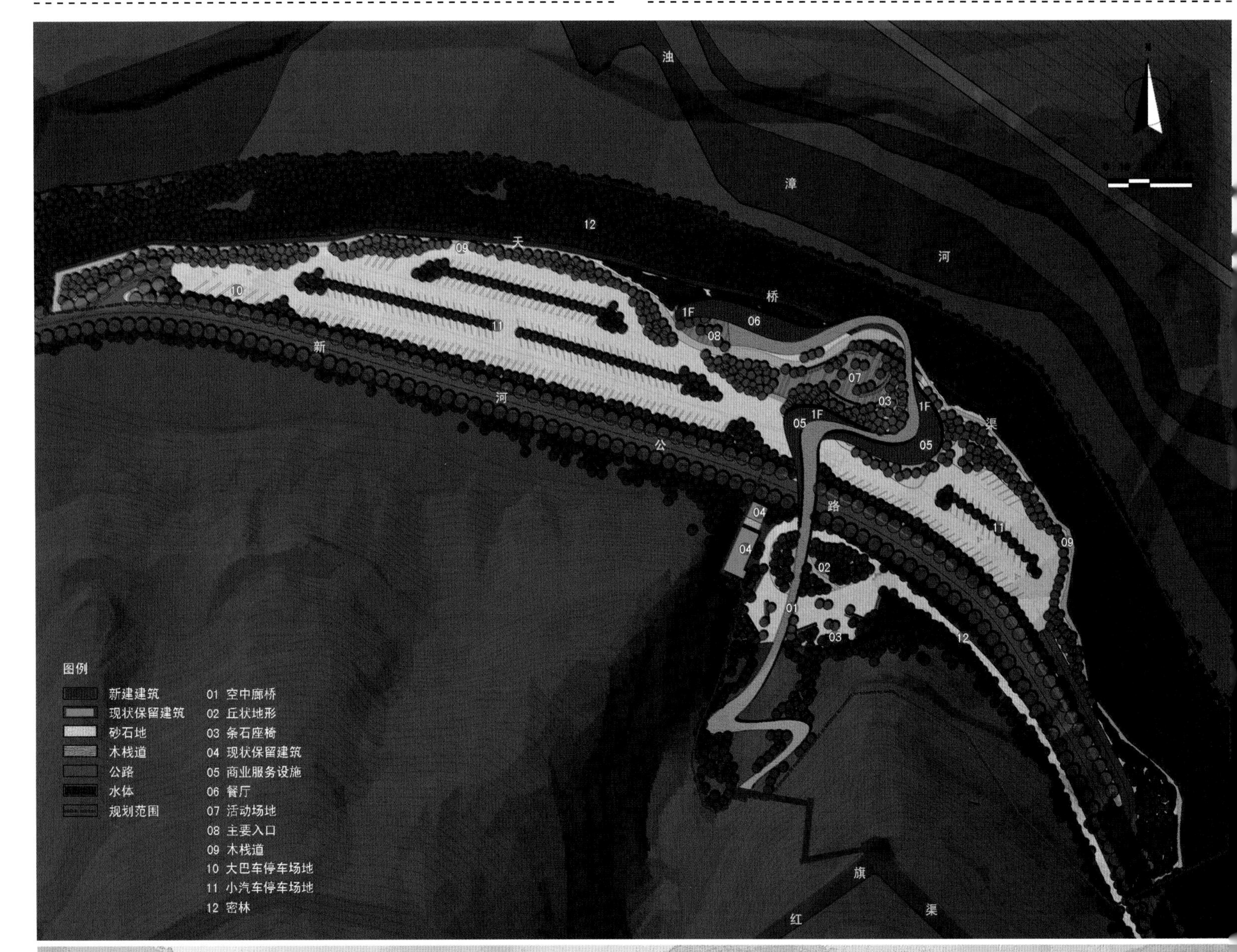

江苏省睢宁县地藏寺建筑概念及周边景观设计方案

项目地点：江苏省 睢宁市
项目规模：8公顷
项目类别：建筑设计、景观设计
委 托 方：睢宁县规划局

Architectural Concept and Surrounding Landscape Architecture Scheme of Dizang Temple, County of Suining, Jiangsu Province

Location: Suining City, Jiangsu Province
Area (size): $8hm^2$
Project Type: Architectural Design, Landscape Architecture
Client: Suining County Planning Bureau

九江县文化艺术中心建筑工程设计

项目地点： 江西省 九江县
项目规模： 25000平方米
项目类别： 建筑设计
委 托 方： 九江县建设局

Architectural Engineering Design of Culture and Art Center, County of Jiujiang

Location: County of Jiujiang, Jiangxi Province
Area (size): 25,000m^2

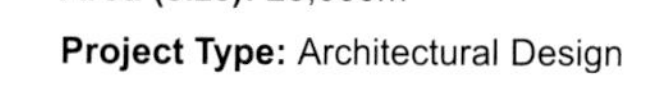

Project Type: Architectural Design
Client: Construction Bureau of the County of Jiujiang

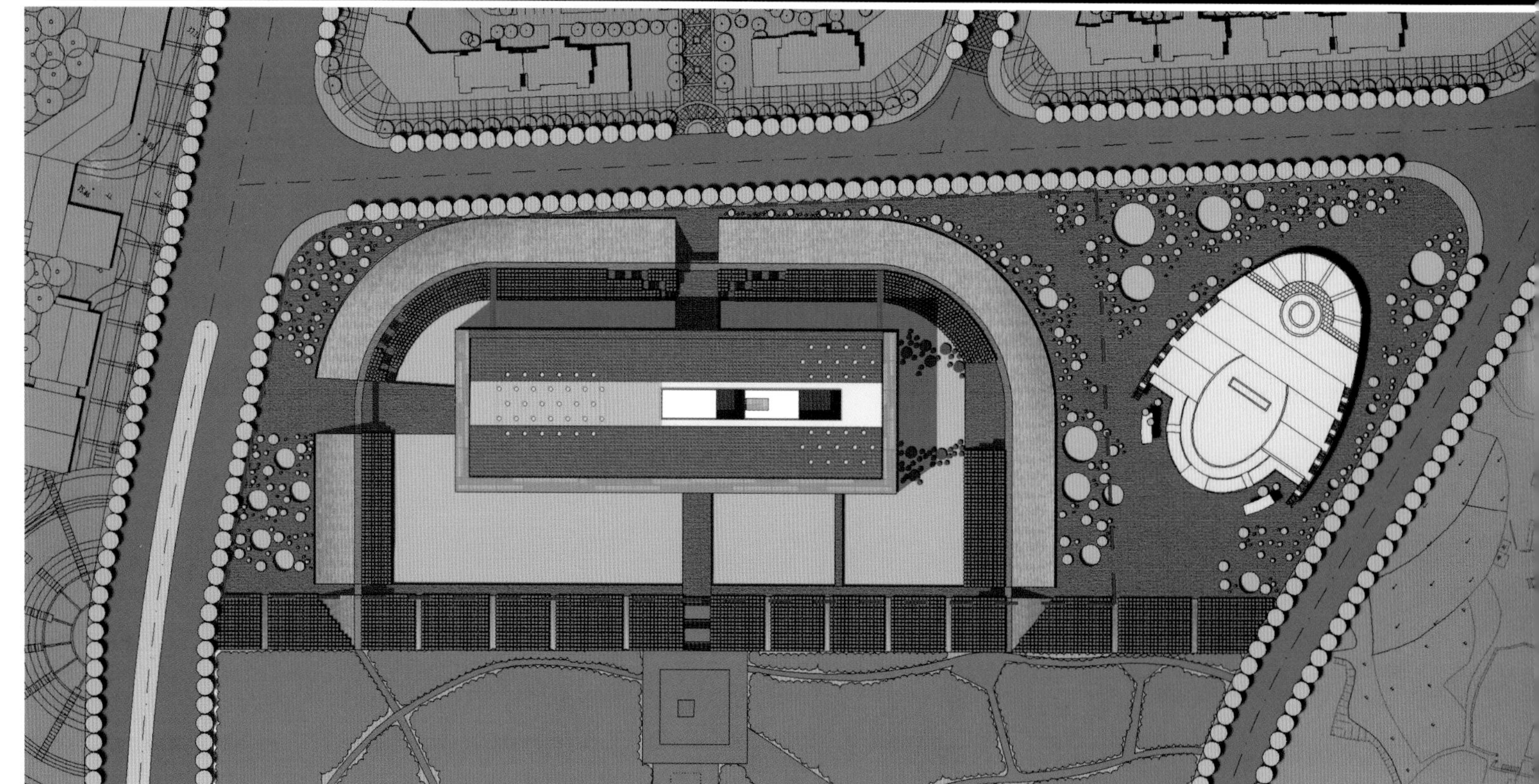

北京环渤海高端总部基地

项目地点：北京市 通州区 台湖镇
项目规模：5.74公顷（研究范围17.21公顷）
项目类别：城市设计、景观设计
委 托 方：北京市规划委员会通州分局
北京市通州区台湖高端总部基地建设管理委员会

High-end Headquarters Base in the Bohai Sea, Beijing

Location: Town of Taihu, Tongzhou District, Beijing
Area (size): 5.74hm^2 (research area: 17.21hm^2)
Project Type: Urban Design, Landscape Architecture
Client: Tongzhou Branch, Beijing Municipal Commission of Urban Planning;
Commission of Construction and Administration of Taihu High-end Headquarters Base, Tongzhou District, Beijing

开远市体育中心建筑工程设计

项目地点：云南省 红河州 开远市

项目规模：6万平方米

项目类别：建筑设计

委 托 方：开远市建设局

Architectural Engineering Design of Sports Center, Kaiyuan City

Location: Kaiyuan City, Honghe Hani & Yi Autonomous Prefecture, Yunan Province

Area (size): 60,000m^2

Project Type: Architecture Design

Client: Construction Bureau of Kaiyuan City

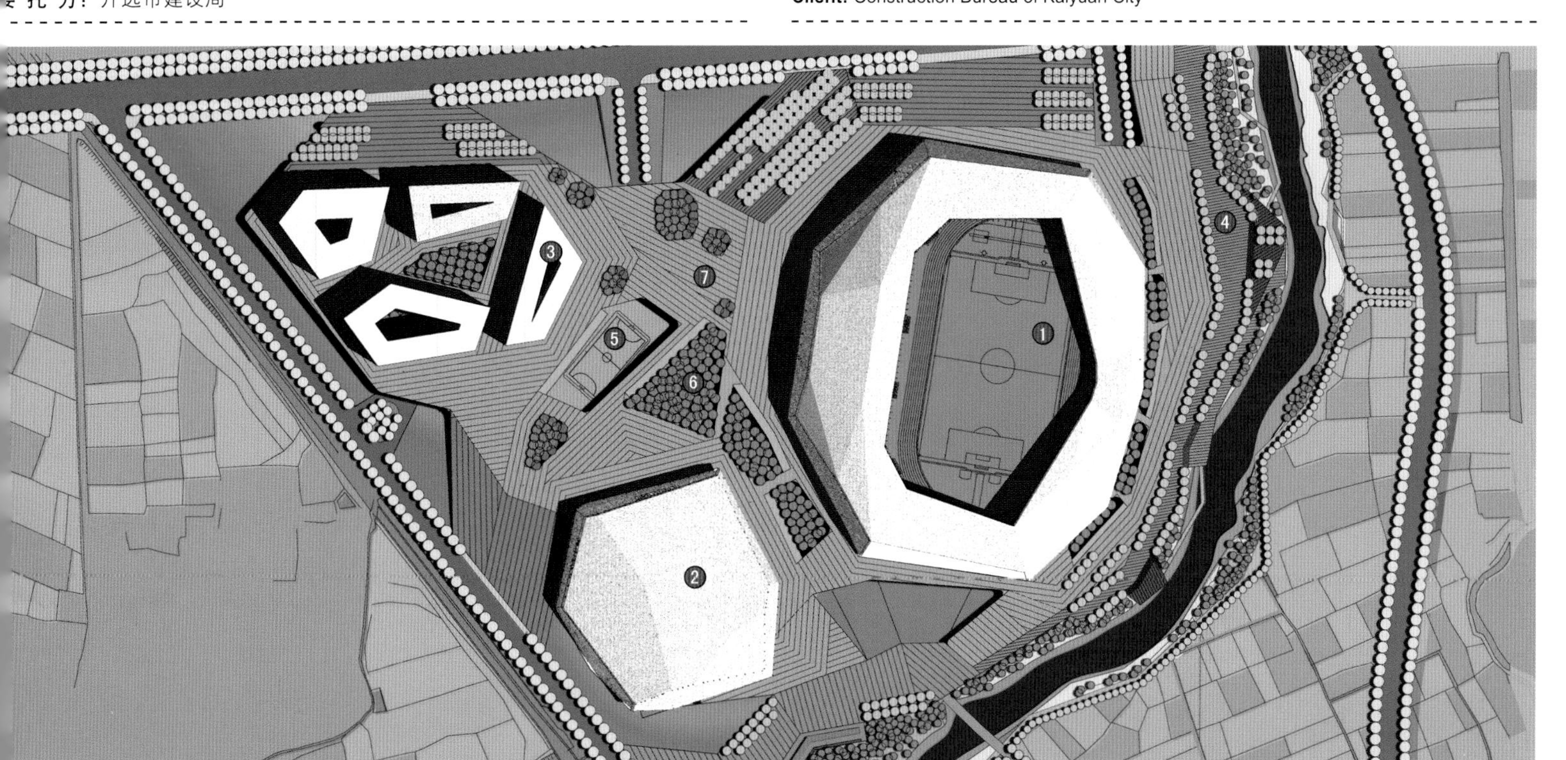

韶山藏书楼项目规划设计

项目地点：昆明 新知

项目规模：40公顷（建筑：9.7万平方米）

项目类别：规划、建筑设计

委 托 方：昆明新知集团有限公司

Planning Design of Shaoshan Library Scheme

Location: Xinzhi, Kunming City

Area (size): $40hm^2$ (architecture area: $97,000m^2$)

Project Type: Planning and Architecture Design

Client: Kunming Xinzhi Group Co., Ltd.

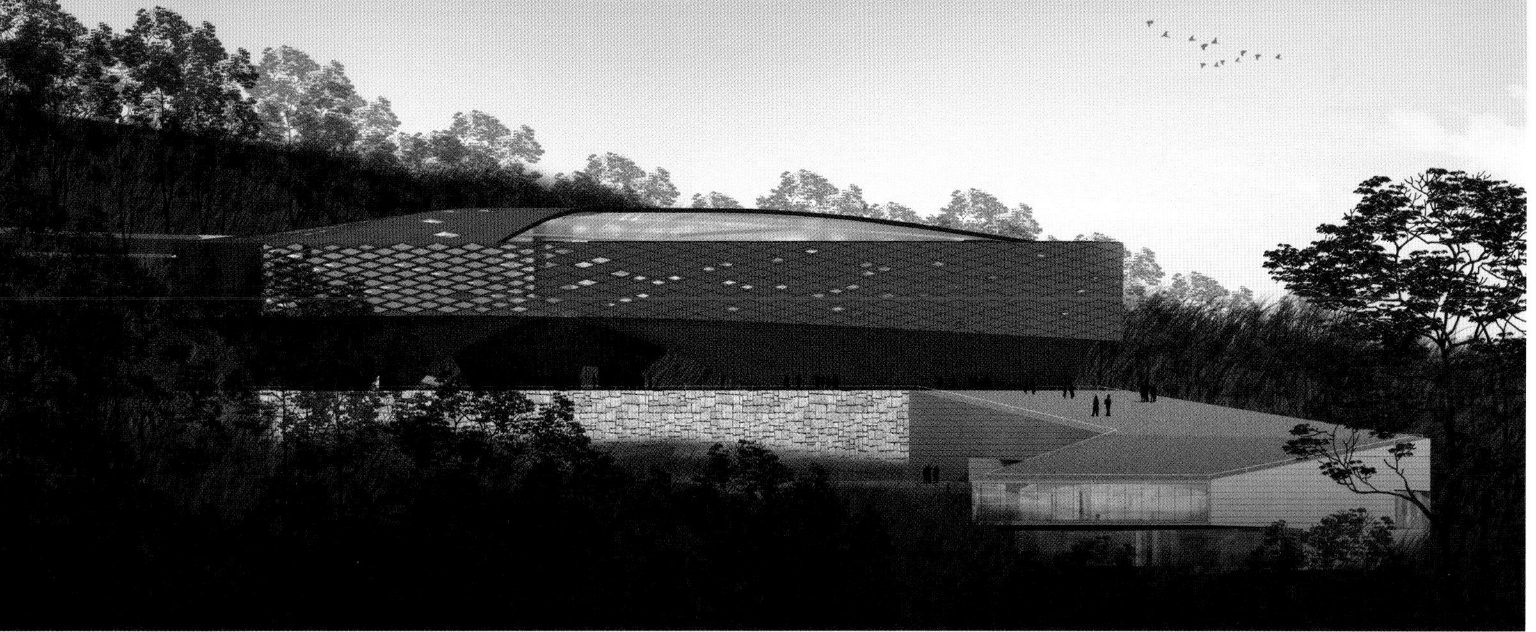

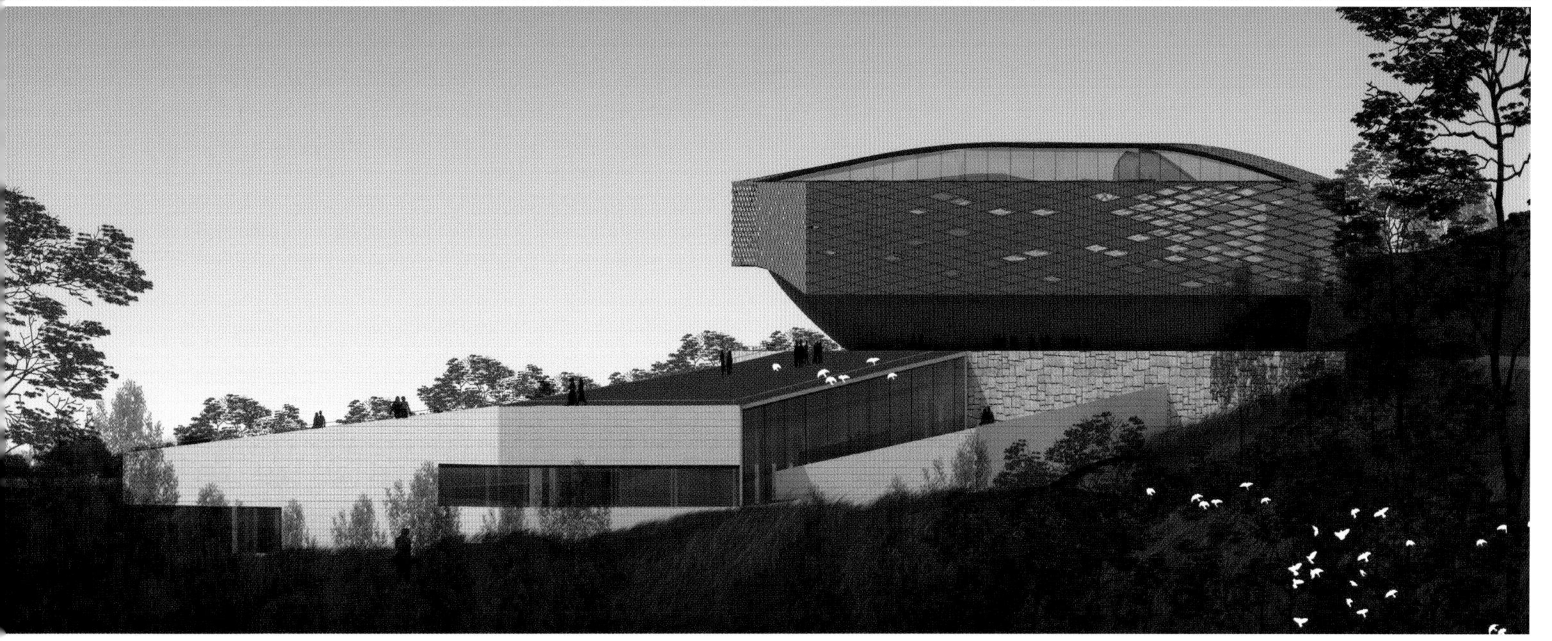

六盘水市中心区荷城古镇改造及周边用地修建性说细规划

项目地点：贵州省 六盘水市

项目规模：54公顷

项目类别：修建性详细规划

委 托 方：六盘水经济开发区管委会开发总公司

Redevelopment of Old Dutch Town, Liupanshui CBD, and rezoning and masterplanning of surrounding area

Location: Liupanshui city, Guizhou Province

Area (size): 54hm^2

Project Type: Masterplanning

Client: Liupanshui Economic Development Zone Development Corporation

琢山濃翳水無波
四望城浮一葉荷
碧伞青溪荷田潴
瑜彩流光萬庭荷

琢山濃翳水無波
四望城浮一葉荷
碧傘青溪荷田漾
瑜彩流光萬庭荷

Heineken

Barbecue
Booth

Boys&Girls
CLUB
LAMAISON
montreal

海南屯昌城市中心区城市设计

项目地点：海南省 屯昌县

项目规模：144公顷

项目类别：城市设计

委 托 方：屯昌县住房和城乡建设局

Urban Design of Tunchang city centre, Hainan

Location: Tunchang city centre, Hainan Province

Area (size): $144hm^2$

Project Type: Urban Design

Client: Department of Housing and Public Works, Tunchang

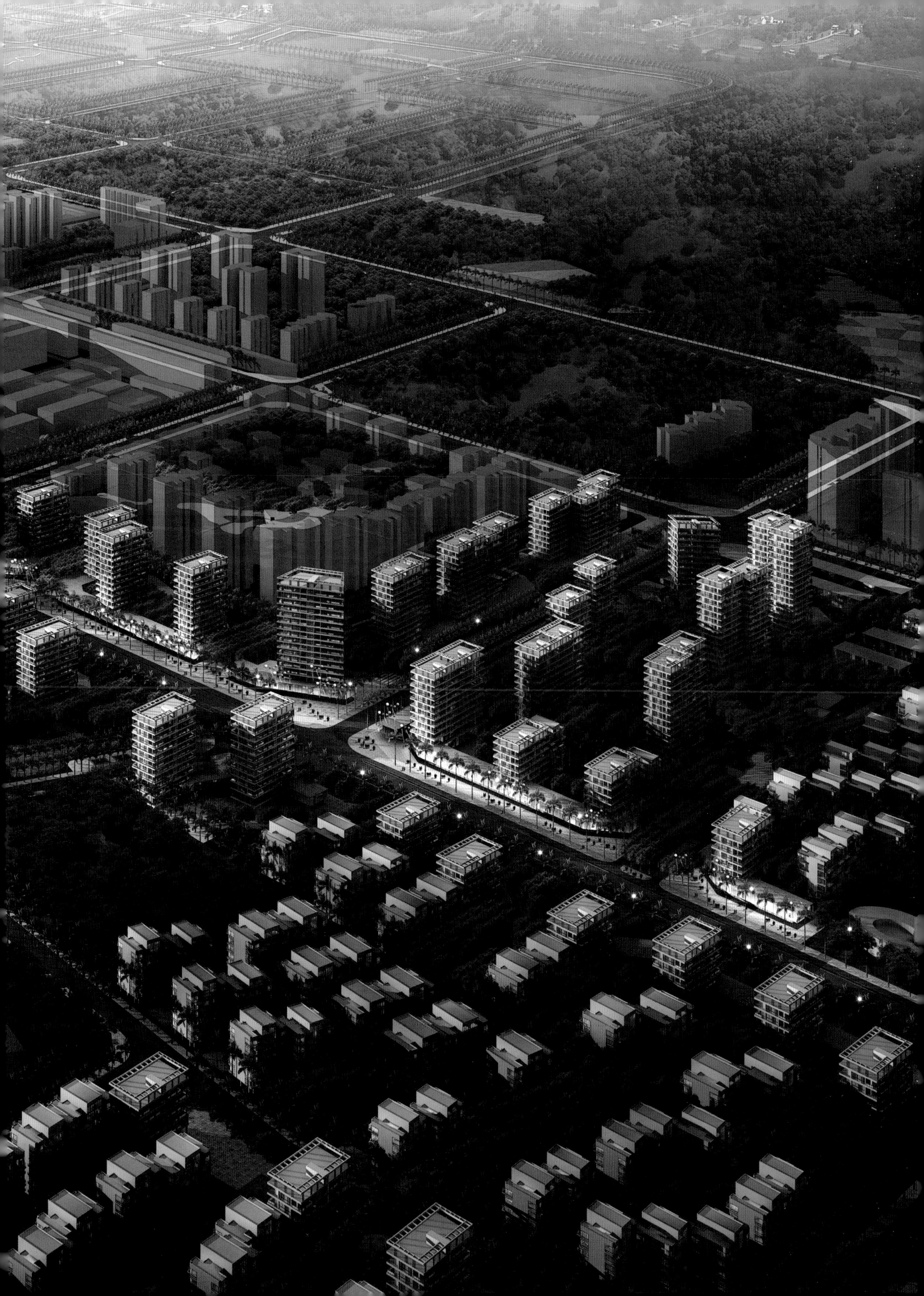

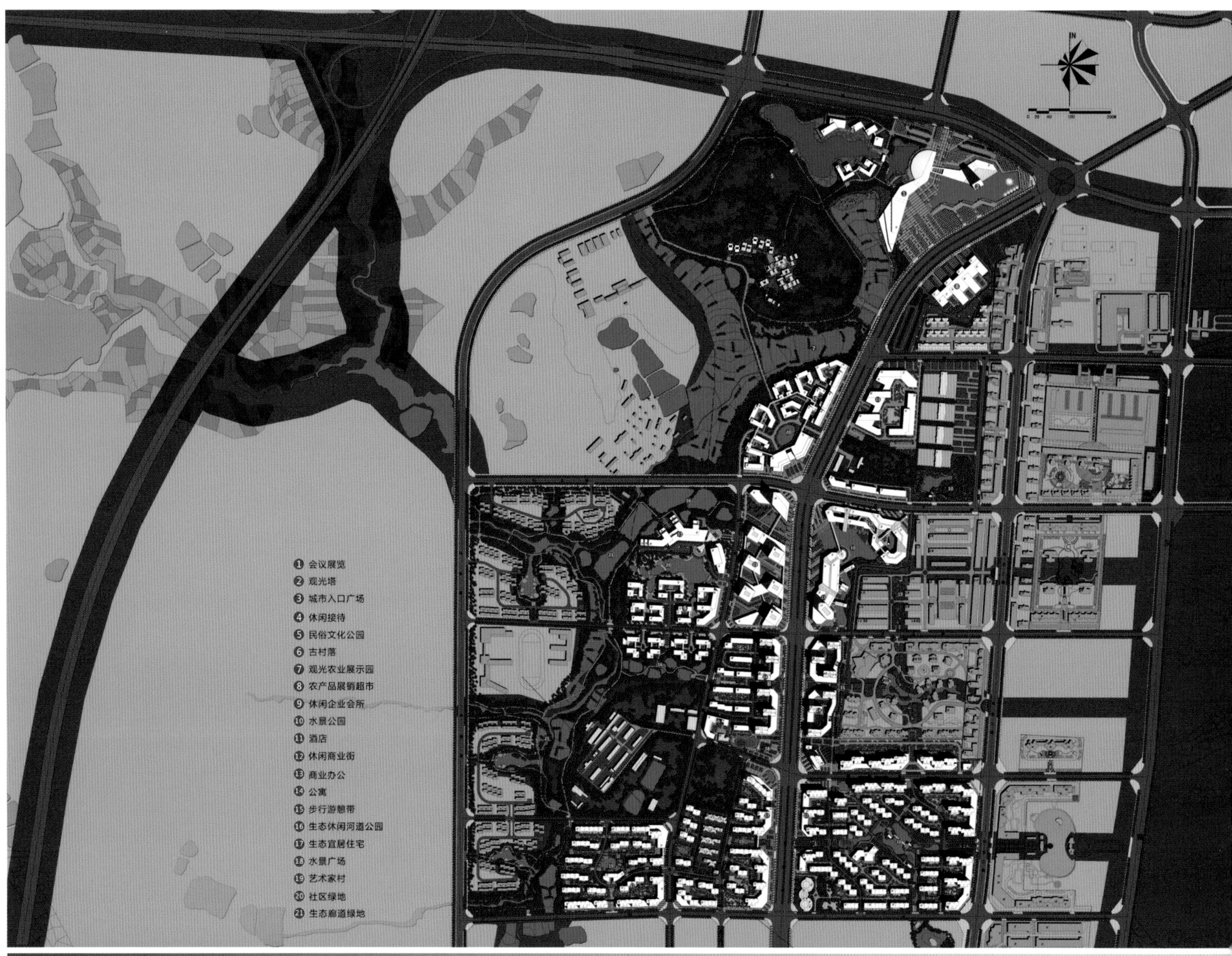
N
1 会议展览
2 观光塔
3 城市入口广场
4 休闲接待
5 民俗文化公园
6 古村落
7 观光农业展示园
8 农产品展销超市
9 休闲企业会所
10 水景公园
11 酒店
12 休闲商业街
13 商业办公
14 公寓
15 步行游憩带
16 生态休闲河道公园
17 生态宜居住宅
18 水景广场
19 艺术家村
20 社区绿地
21 生态廊道绿地

Dior
Dior

山海关城区总体城市设计

项目地点：河北省 山海关市
项目规模：22.73平方千米
项目类别：城市设计
委 托 方：山海关规划局

Urban Design of Shanhaiguan Metropolitan Area

Location: Shanhaiguan City, Hubei Province
Area (size): 22.73km^2
Project Type: Urban Design
Client: Department of Planning, Shanhaiguan

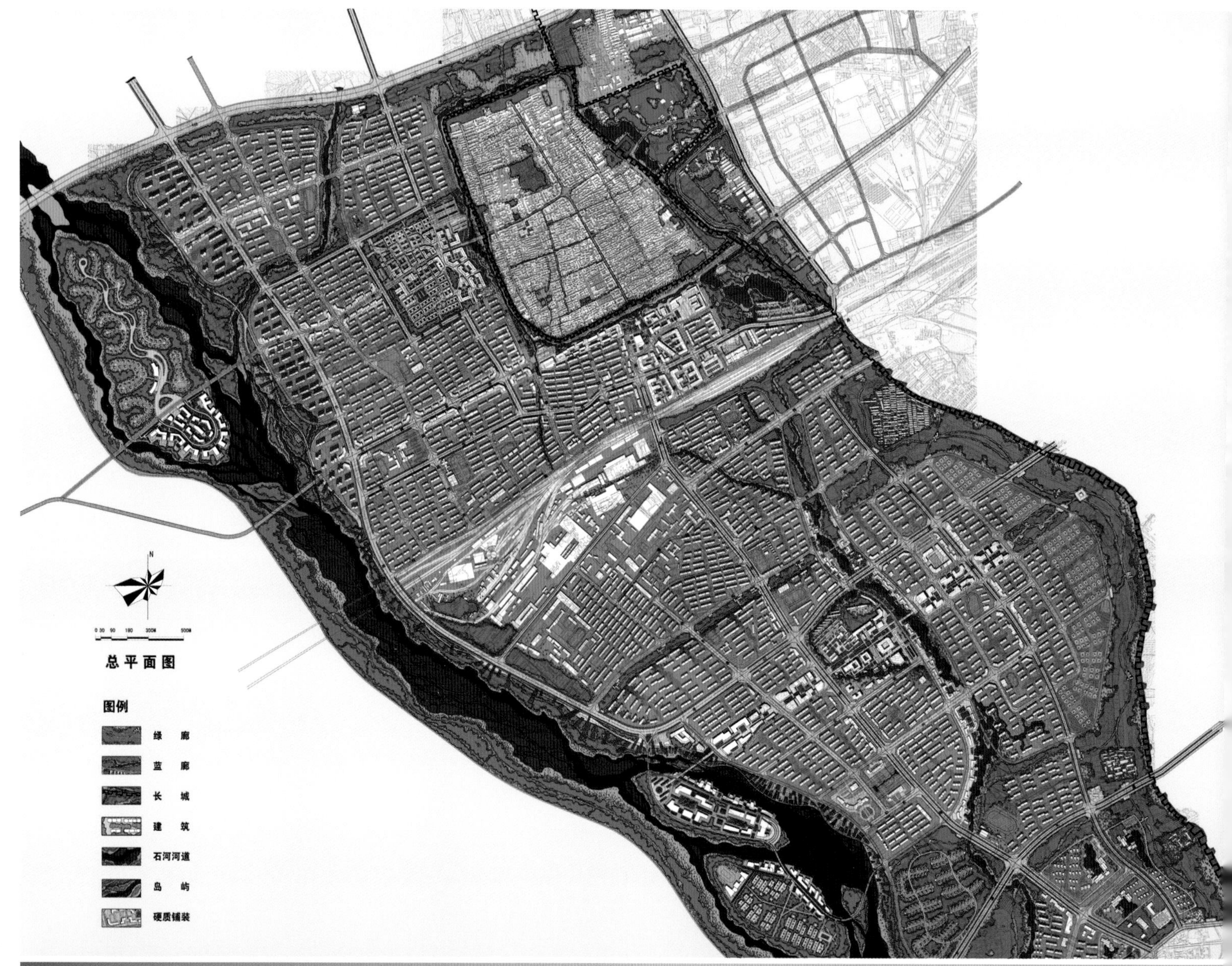

武汉市中国光谷依托邦城市设计

项目地点：湖北省 武汉市 江夏区
项目规模：22平方千米
项目类别：城市设计
委 托 方：湖北大都置业有限公司

Urban Design of China Optics Valley, Wuhan City

Location: Jiangxia District, Wuhan City, Hubei Province
Area (size): $22km^2$
Project Type: Urban Design
Client: Hubei Metro Properties Limited

N
0 100 200 500 1000M
Legend
Villa
Countryard Villa
Multistory Building
High Building
Auxiliary Service Building
The Youth High-tech Industrial Park
Urban Commercial Building
Planning Road
Primary Corridor
Subsidiary Corridor
Branch Corridor
Water Body
Hill
Existing Village

Dior

株洲市轨道科技城城市景观风貌规划暨重点区域城市设计

项目地点：湖南省 株洲市

项目规模：30公顷

项目类别：城市设计、景观设计

委 托 方：株洲市规划局

Urban Design of key zones and landscape Design and planning of Track Technology Park, Zhuzhou City

Location: Zhuzhou City, Hunan Province

Area (size): 30hm^2

Project Type: Urban Design, landscape Design

Client: Zhuzhou City Planning Department

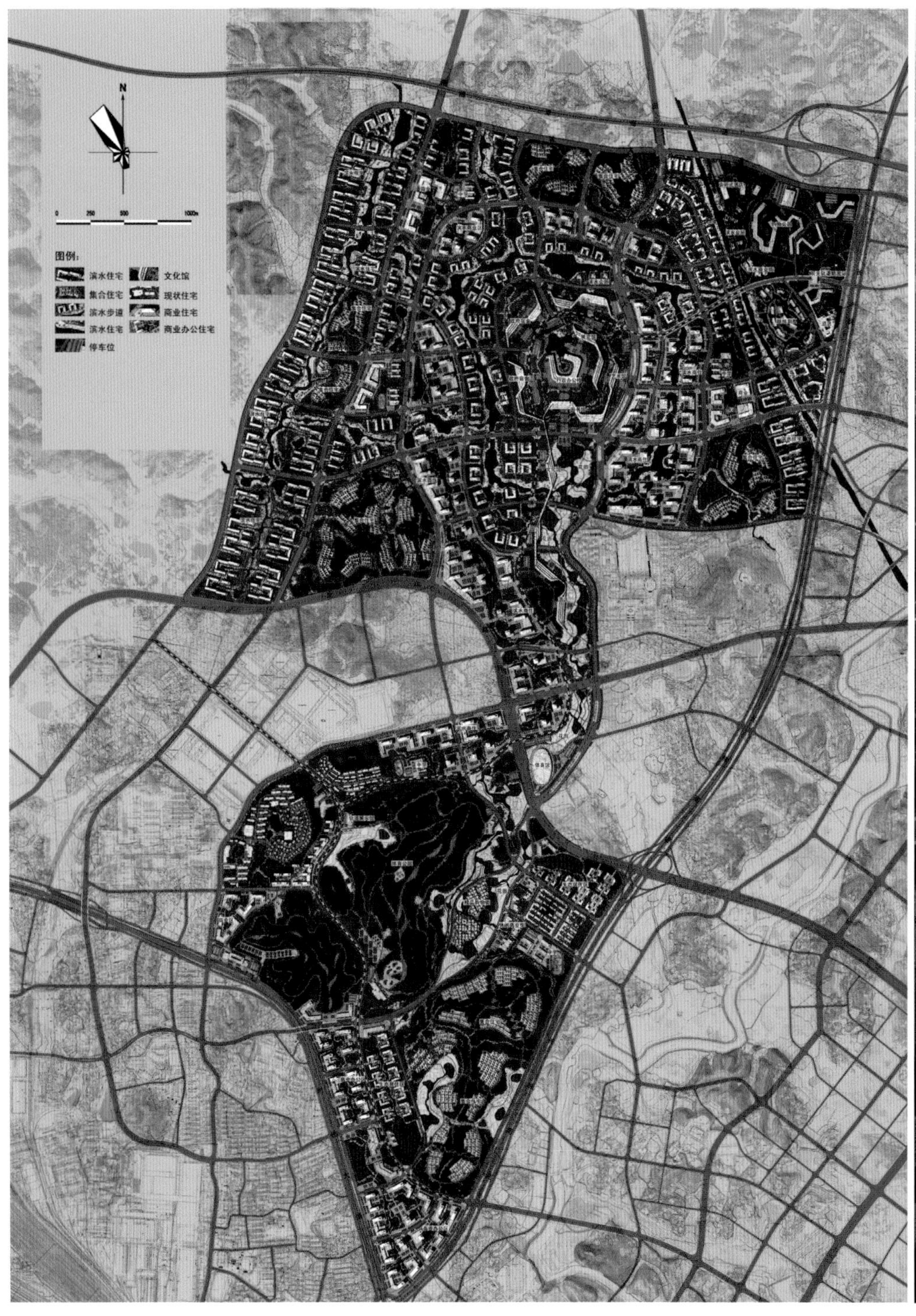

重庆市江津区北部新城城市设计
项目地点：重庆市 江津区
项目规模：30平方千米
项目类别：城市设计
委 托 方：重庆市规划研究中心
Urban Design of northern sector, Jiangjin District, Chongqing
Location: Jiangjin District, Chongqing City
Area (size): 30km2
Project Type: Urban Design
Client: Chongqing Urban Planning Research Centre

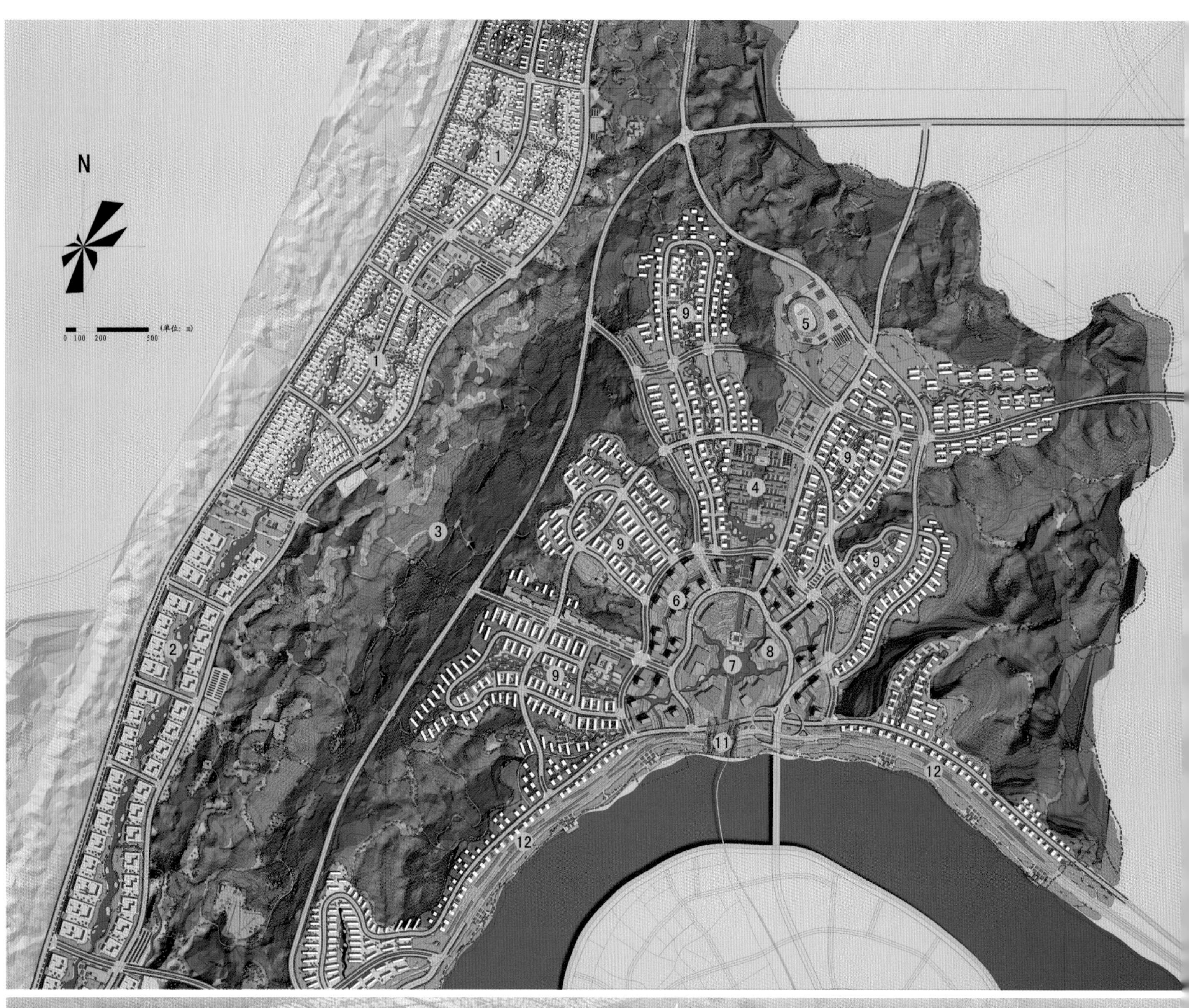
N
0 100 200 500
(单位：m)
1
1
2
3
4
5
6
7
8
9
9
9
9
9
11
12
12

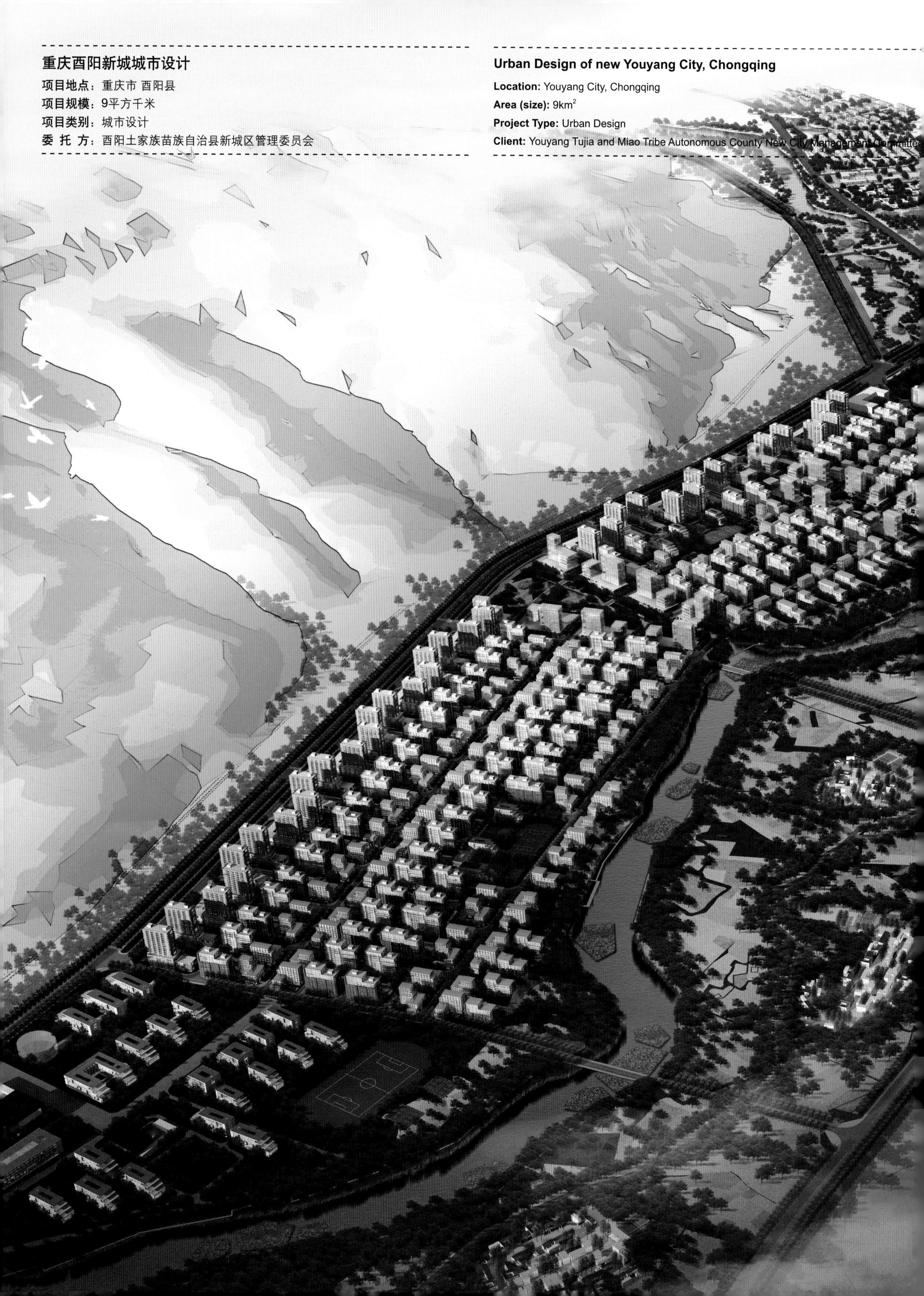

重庆酉阳新城城市设计

项目地点：重庆市 酉阳县

项目规模：9平方千米

项目类别：城市设计

委 托 方：酉阳土家族苗族自治县新城区管理委员会

Urban Design of new Youyang City, Chongqing

Location: Youyang City, Chongqing

Area (size): $9km^2$

Project Type: Urban Design

Client: Youyang Tujia and Miao Tribe Autonomous County New City Management Committe

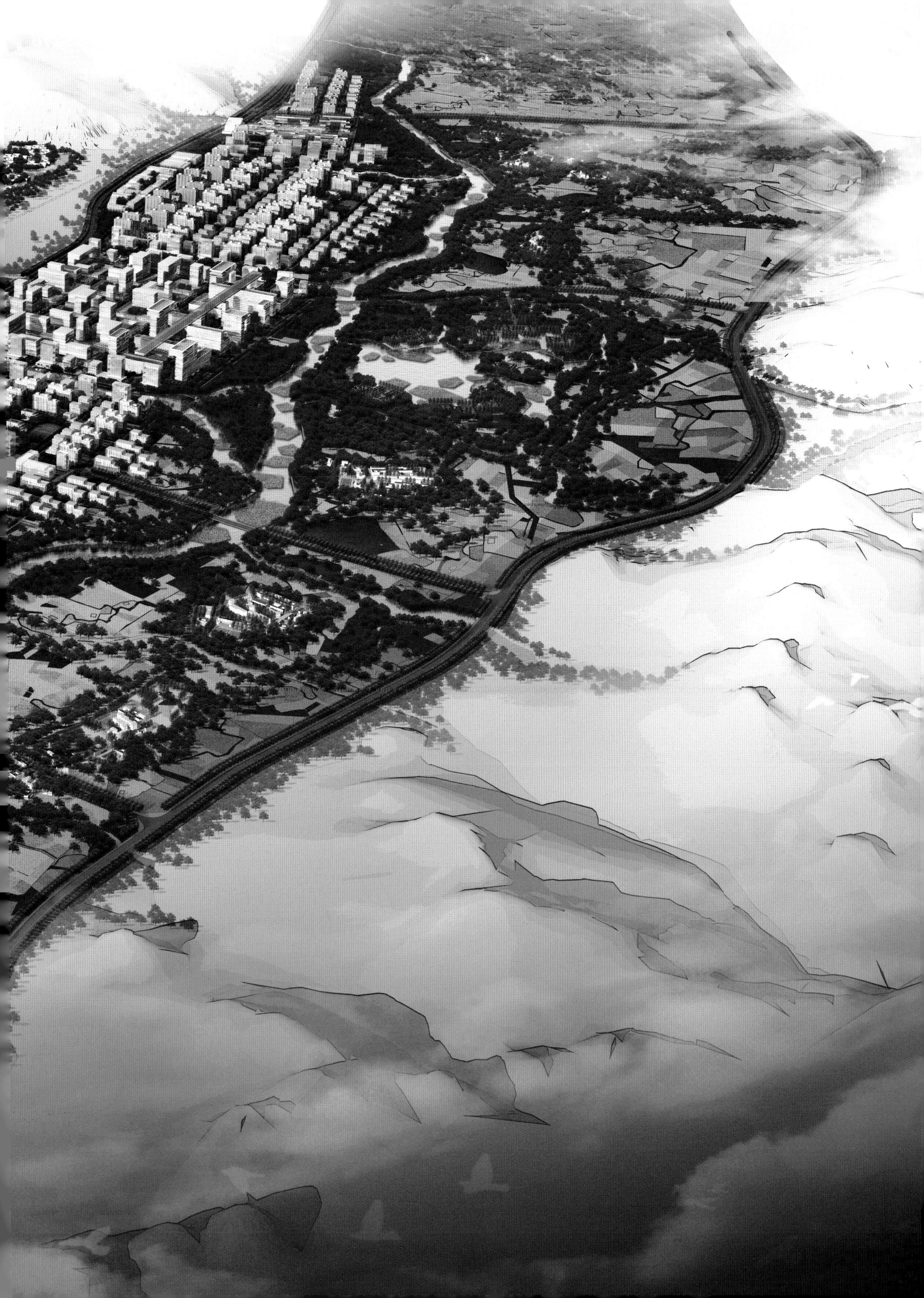

N
0 50 100 400M
交易市场
火车站
公交总站
小学
滨水酒店
体疗养村落
花田
三甲医院
渤海小镇住区
城市公园
现状保留商业
渝
怀
铁
路
步行街区
完全中学
博物馆
滨水木栈道
观景平台
文化馆
精品展示中心
商业金融
行政中心
歌剧院
古树公园
消防站
小学
滨水酒店
体疗养村落
步行街区
混合商业
服务业
滨水步道
花田
城市公园
自行车道
教育科研
体疗养村落
商住
教育科研
小学
滨水酒店
龙潭一中
图例：
建筑
水泡
林地
花田
道路
自行车道
步行栈道
停车场

江苏高淳科技创业特别社区城市设计

项目地点：江苏省 高淳县

项目规模：4.9平方千米

项目类别：城市设计

委 托 方：高淳县住房与建设委员会

Urban Design of Innovation Technology Special Zone, Jiangsu, Gaochun

Location: Gaochun, Jiangsu Province

Area (size): 4.9km^2

Project Type: Urban Design

Client: Gaochun Housing and Construction Committee

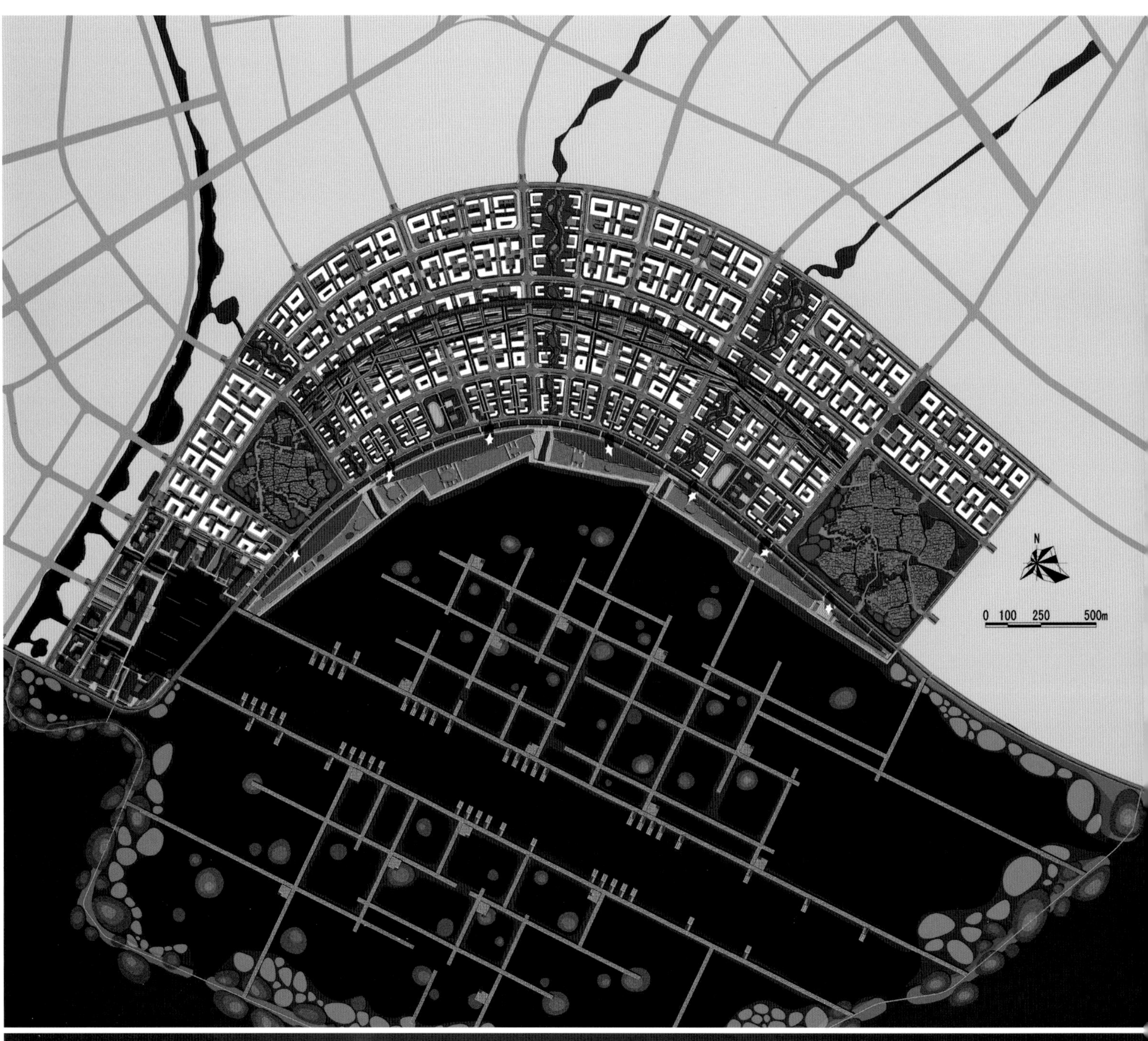
N
0 100 250 500m

陕西省西咸新区渭河生态景观带总体方案设计

项目地点：陕西省 西咸

项目规模：150平方千米

项目类别：城市设计、景观设计

委 托 方：陕西省西咸新区管委会

Ecological landscape framework Design for Wei River, Xixian New Area, Shan

Location: Xixian New Area, Shanxi Province

Area (size): 150km^2

Project Type: Urban Design, landscape Design

Client: Xixianxin District Management Committee

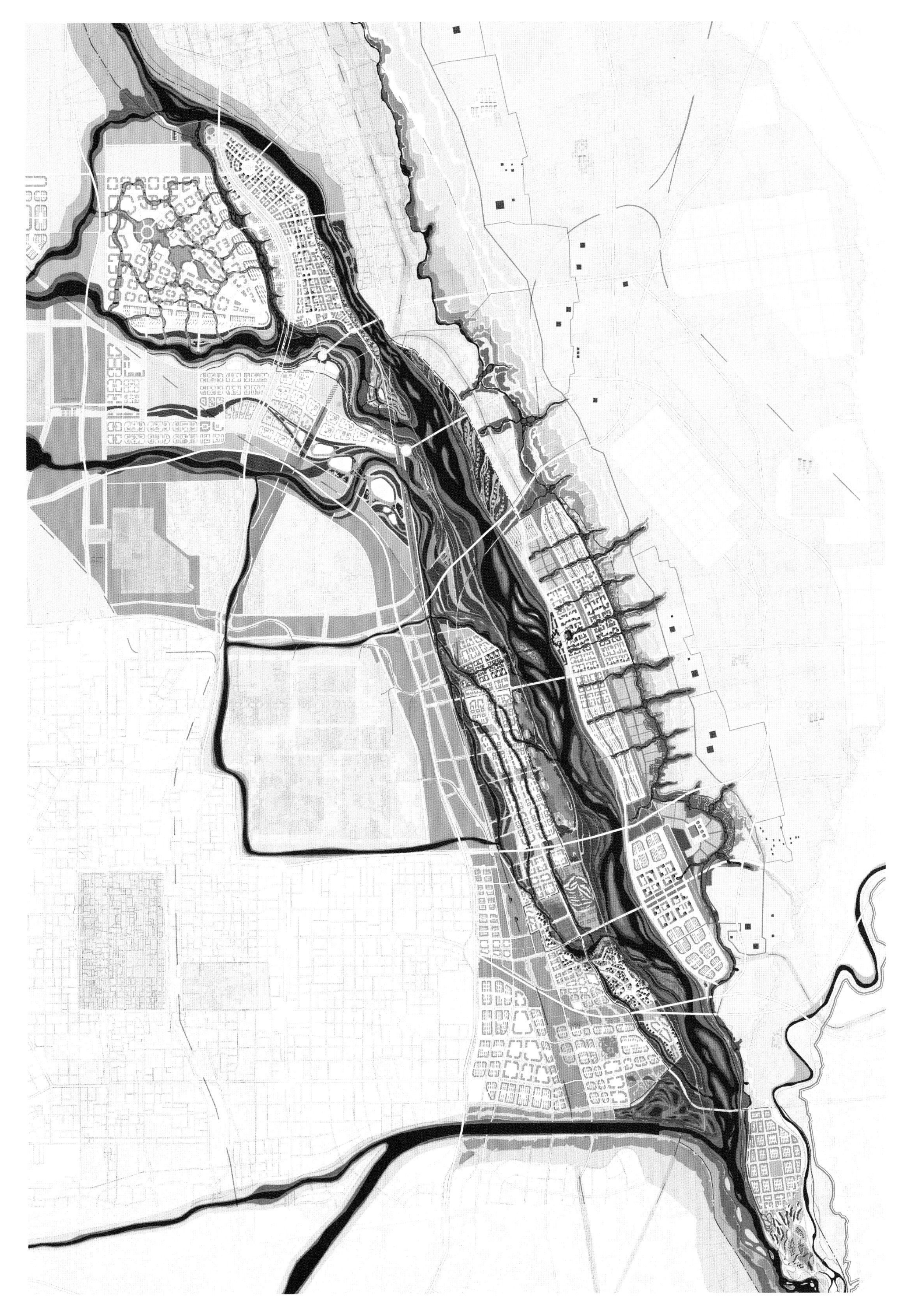

大英县郪江新城城市设计

项目地点：四川省 大英县 郪江新城

项目规模：412公顷

项目类别：城市设计

委 托 方：大英县城乡规划局

Urban Design for New Quarter Qijiang City, Daying

Location: New Quarter Qijiang City, Daying, Sichuan Province

Area (size): 412hm^2

Project Type: Urban Design

Client: Daying Urban and Rural Planning Bureau

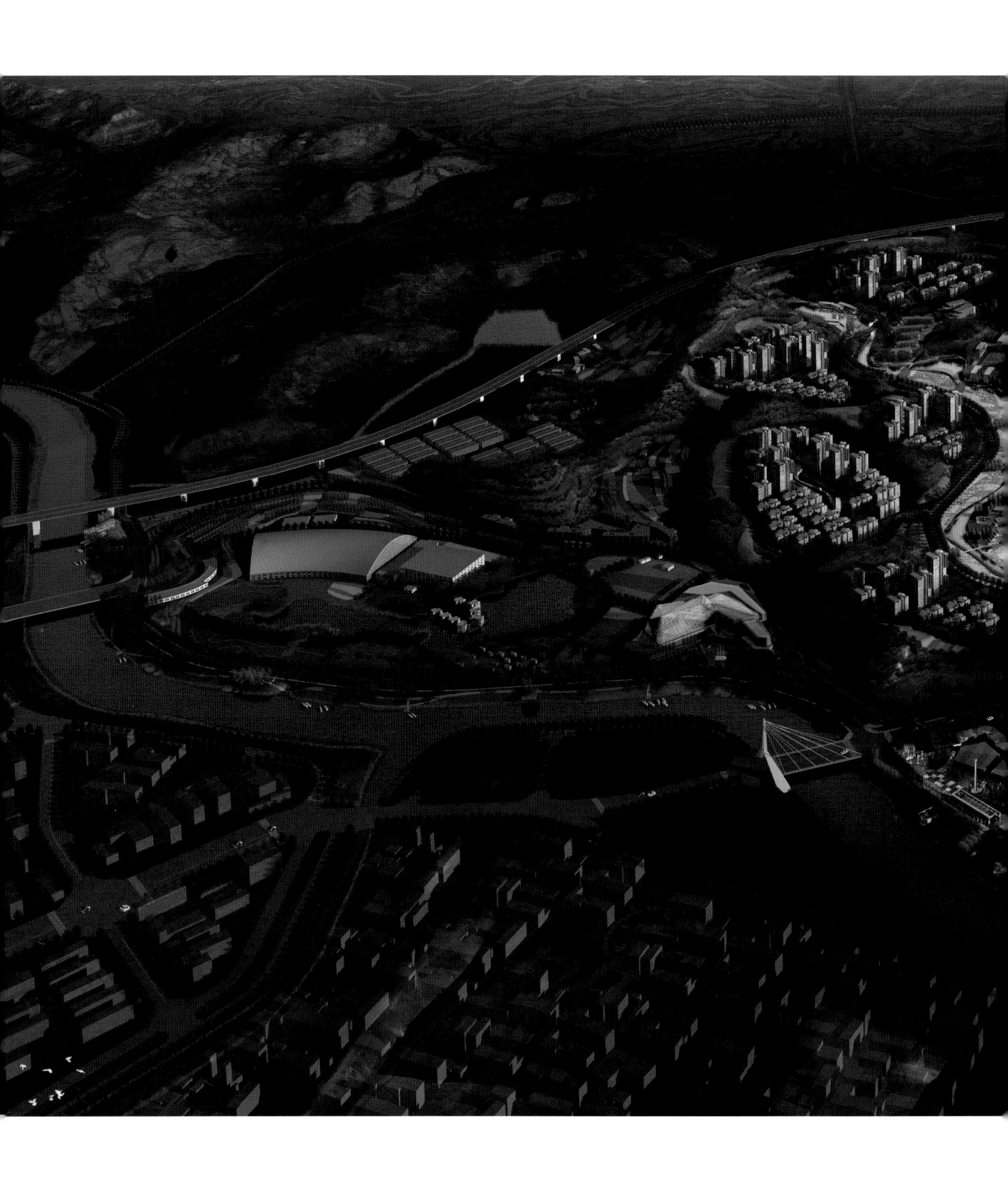

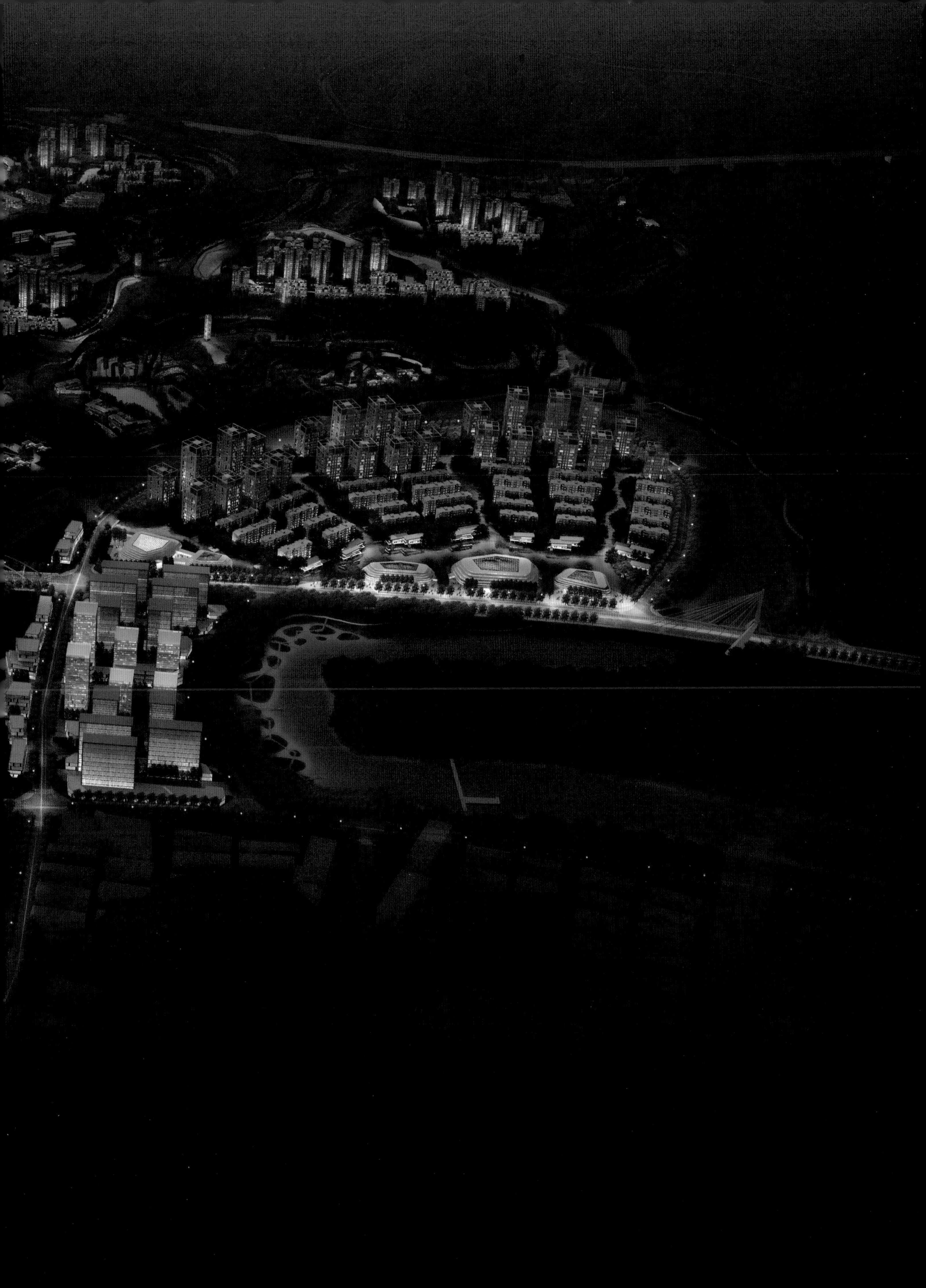

都市农业节点
立体农园
都市农场
都市农业廊道

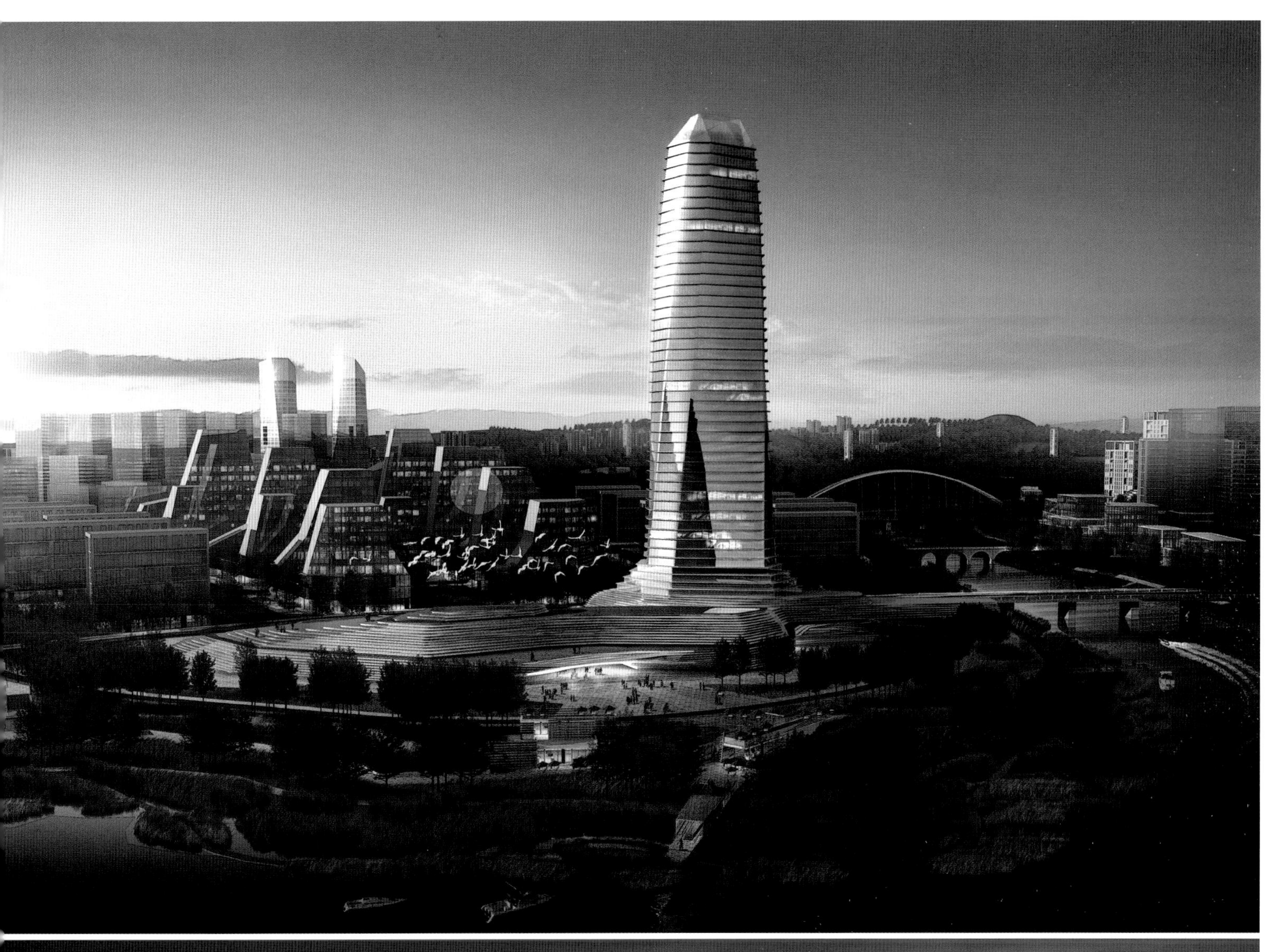

新疆鄯善县城区重点地段城市修建性详细规划

项目地点：新疆 鄯善县
项目规模：300公顷
项目类别：城市设计
委 托 方：新疆鄯善县建设局

Masterplanning of key urban areas in Shanshan, Xinjiang

Location: Shanshan , Xinjiang
Area (size): 300hm^2
Project Type: Urban Design
Client: Shanshan Construction Bureau

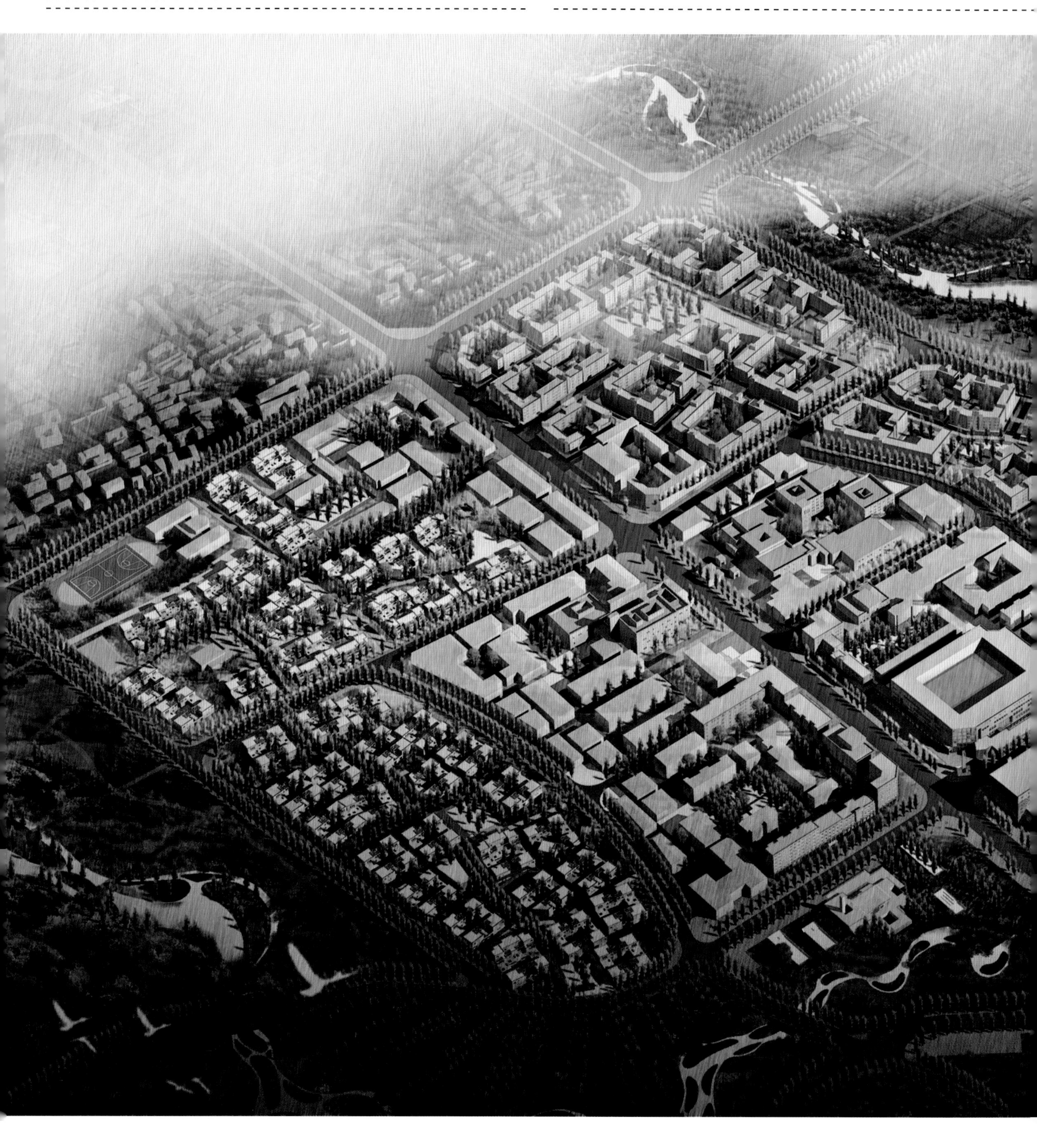

N

DAKOTA
DAKOTA

DAKOTA

PEPER

遵义云顶阳光项目修建性详细规划设计

项目地点：贵州省 遵义市

项目规模：98.37公顷

项目类别：修建性详细规划

委 托 方：盛邦房地产开发有限公司

Masterplan for Wenting Project, Zunyi, Guizhou

Location: Zunyi City, Guizhou Province

Area (size): 98.37hm^2

Project Type: Masterplanning

Client: Sheng Bang Real Estate Development Co., Ltd.

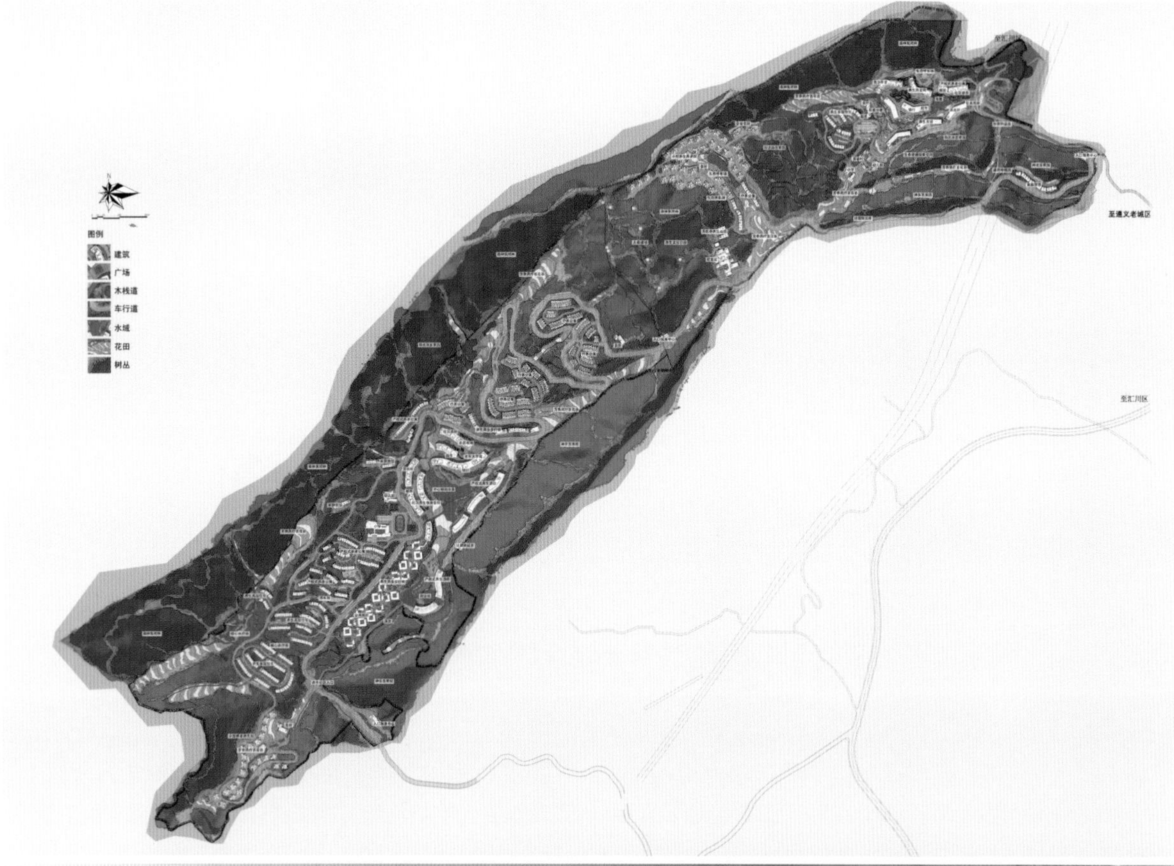

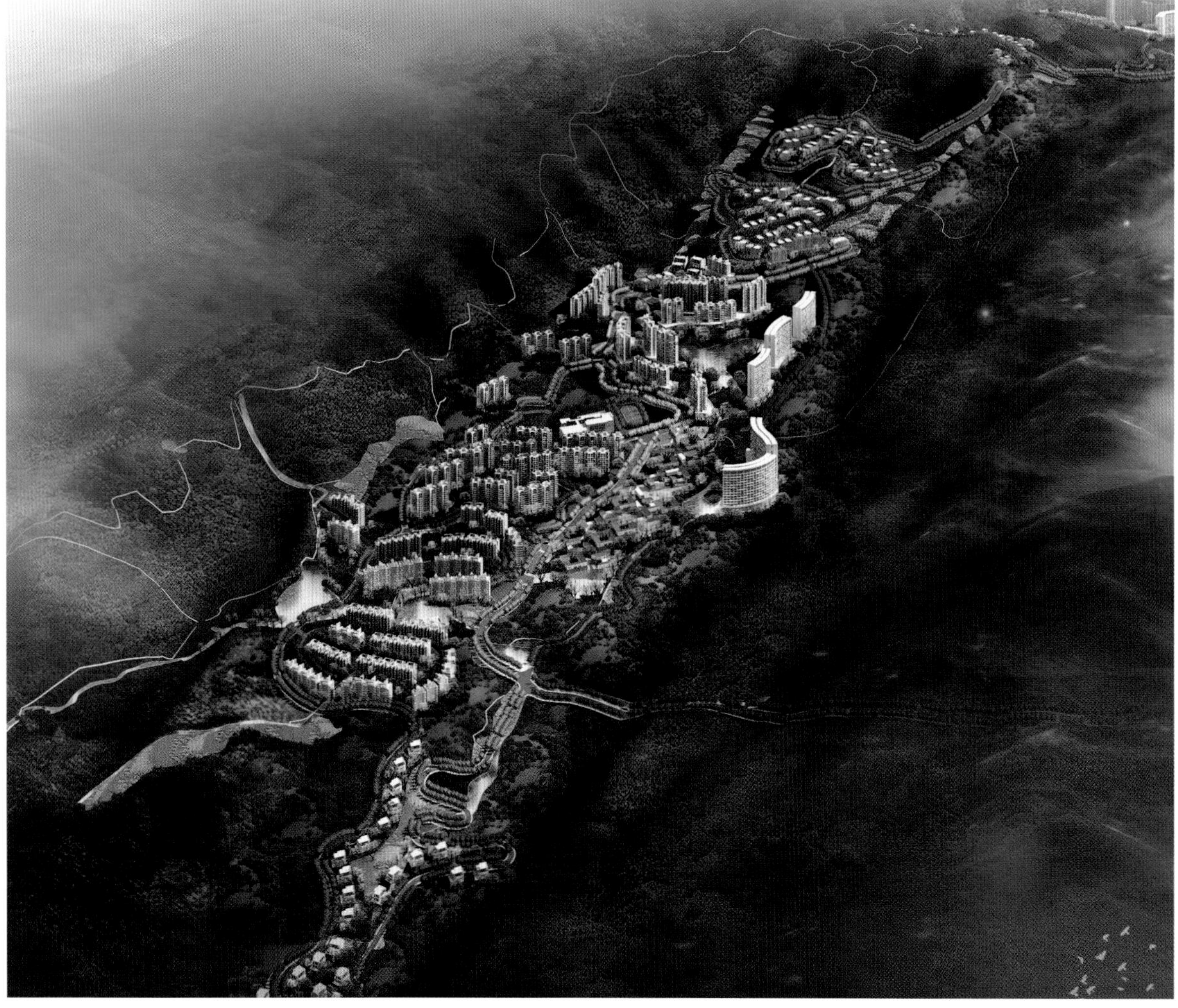

Dior

项目名称：美国明尼阿波利斯密西西比河滨水设计
项目地点：美国 明尼苏达州 明尼阿波利斯市
动画时长：3分30秒
委 托 方：明尼阿波利斯公园与休憩委员会

Project name: The Minneapolis Waterfront Design Concept
Location: Minneapolis, Minnesota, USA
Movie Length: 3min 30sec
Client: The Minneapolis Park & Recreation Board and The Minneapolis Parks Foundation

项目名称：巴西里约热内卢奥林匹克公园设计
项目地点：巴西 里约热内卢
动画时长：3分钟
委 托 方：巴西里约热内卢奥委会

Project name: A 'Carioca' Legacy: Olympic Park Rio 2016
Location: Rio de Janeiro, Brazil
Movie Length: 3min
Client: Rio 2016 Organizing Committee for the Olympic

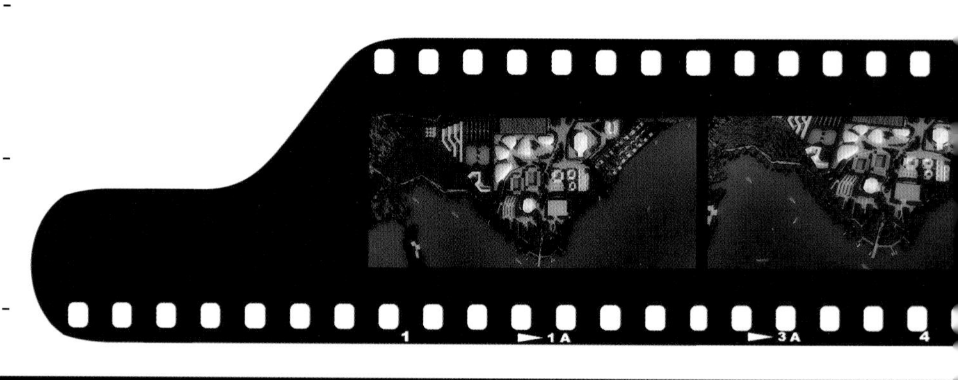

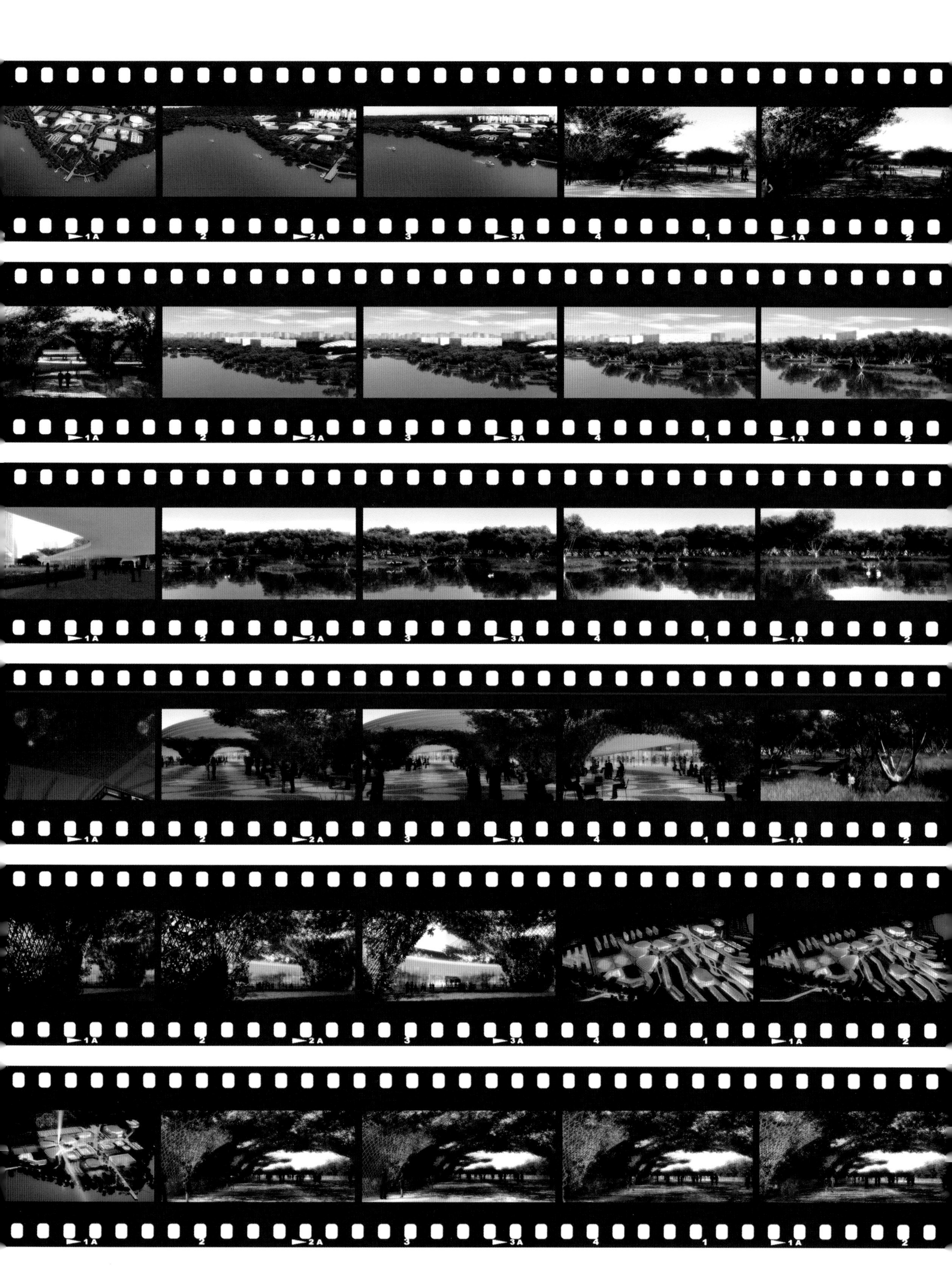

项目名称：北京环渤海高端总部基地城市景观轴线景观设计
项目地点：北京市 通州区 台湖镇
动画时长：3分30秒
委 托 方：北京市规划委员会通州分局，北京市通州区台湖高端总部基地建设管理委员会

Project name: Landscape Design for Beijing Bohai-rim Advanced Business Park Landscape Axis
Location: Tongzhou District, Beijing
Movie Length: 3min 30sec
Client: Beijing Planning Committee, Tongzhou Branch and Beijing Bohai-rim Advanced Business Park Development Management Committee

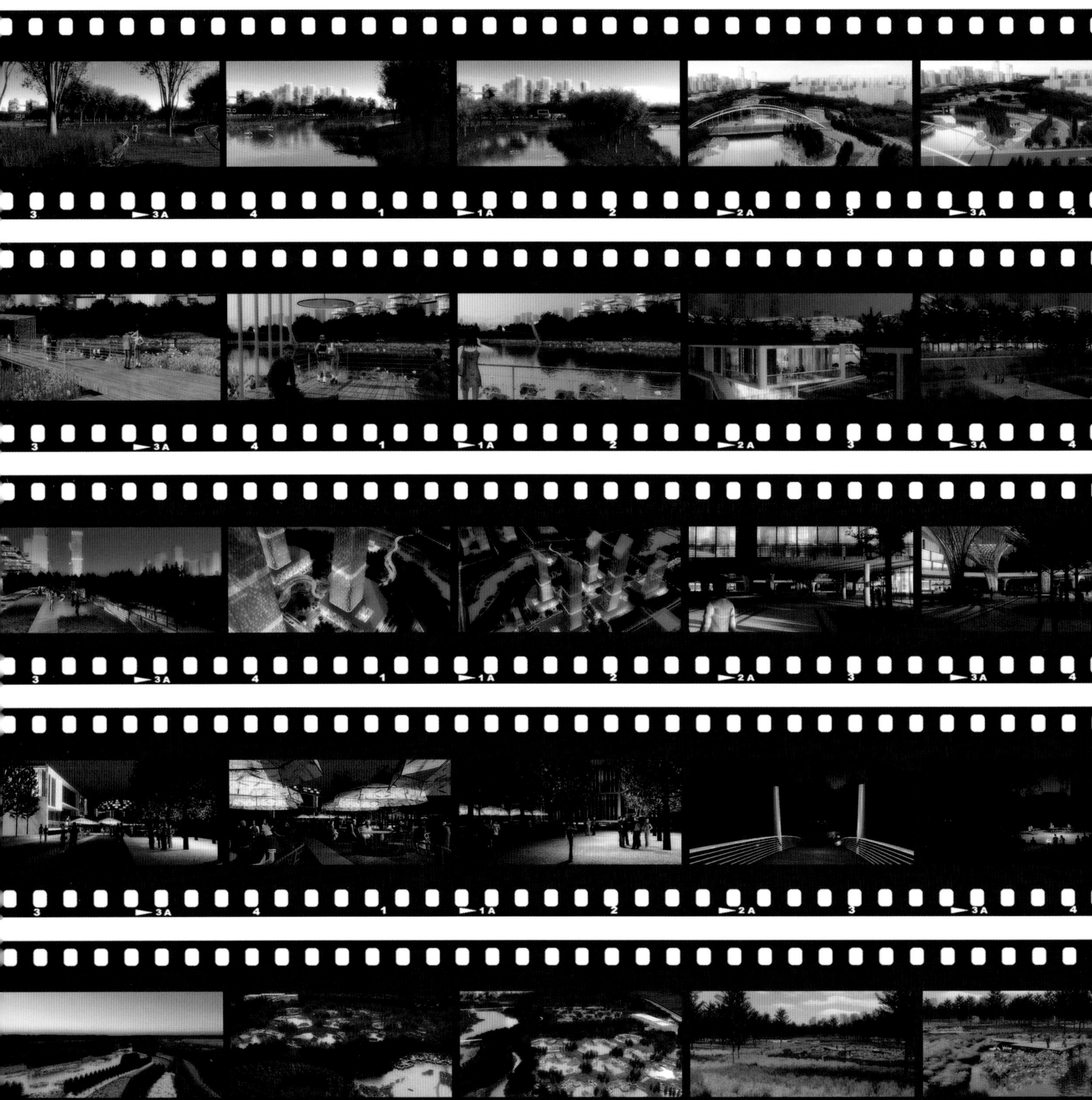

项目名称：广州市天河智慧城生态基础设施暨开放空间研究及天河智慧城生态基础设施规划
项目地点：广州市 天河区
动画时长：5分钟
委 托 方：广州高新技术产业开发区天河科技园管管理委员会

Project name: Ecological Infrastructure and Open Spaces Study of Zhihui City and Ecological Infrastructure Planning for Zhihui City, Tianhe District
Location: Tianhe District, Guangzhou City
Movie Length: 5min
Client: Guangzhou Hi-tech Industrial Development Zone, Tianhe Science Park Management Committee

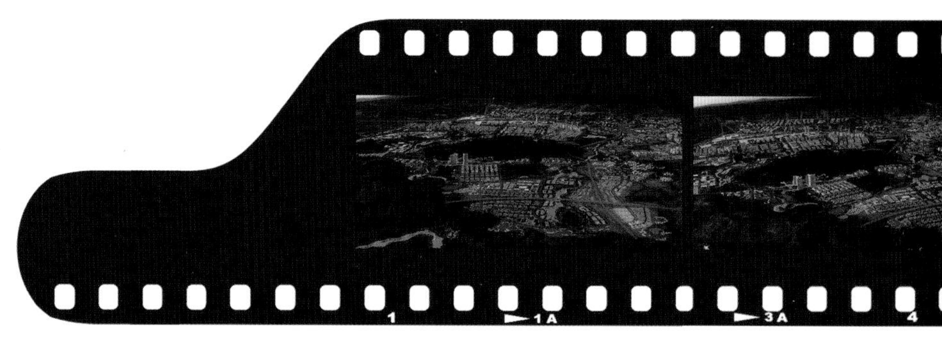

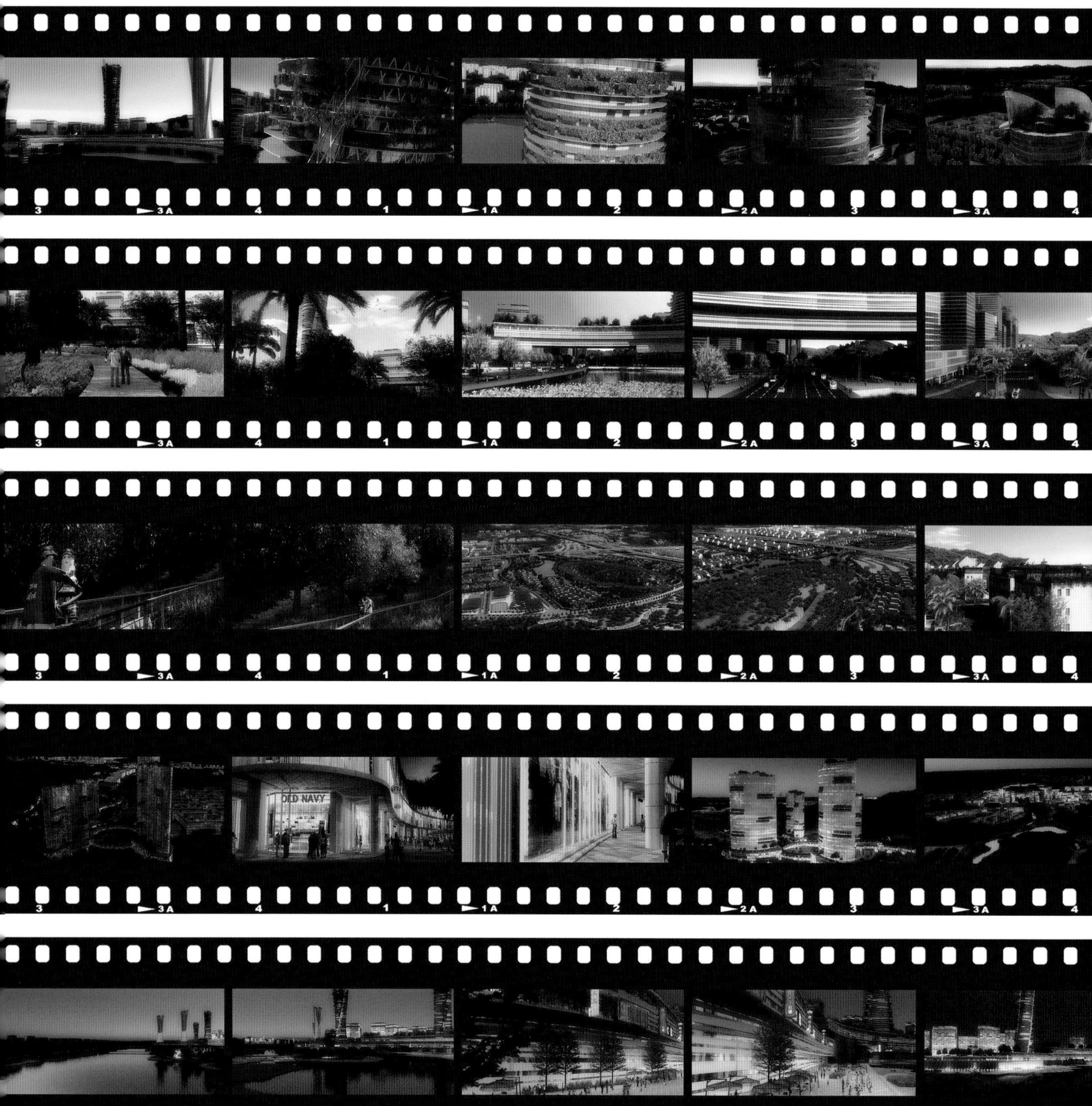

项目名称：芝加哥北格兰特公园设计
项目地点：美国 芝加哥市
动画时长：3分30秒
委 托 方：芝加哥市公园管理局

Project name: The Art Field: The Chicago North Grant Park
Location: Chicago, Illinois, USA
Movie Length: 3min 30sec
Client: Design and Construction Administration Services, Central Park District, Chicago, Illinois

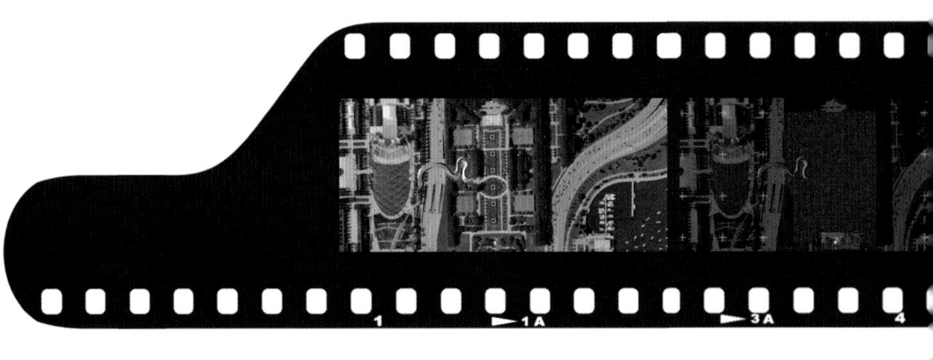

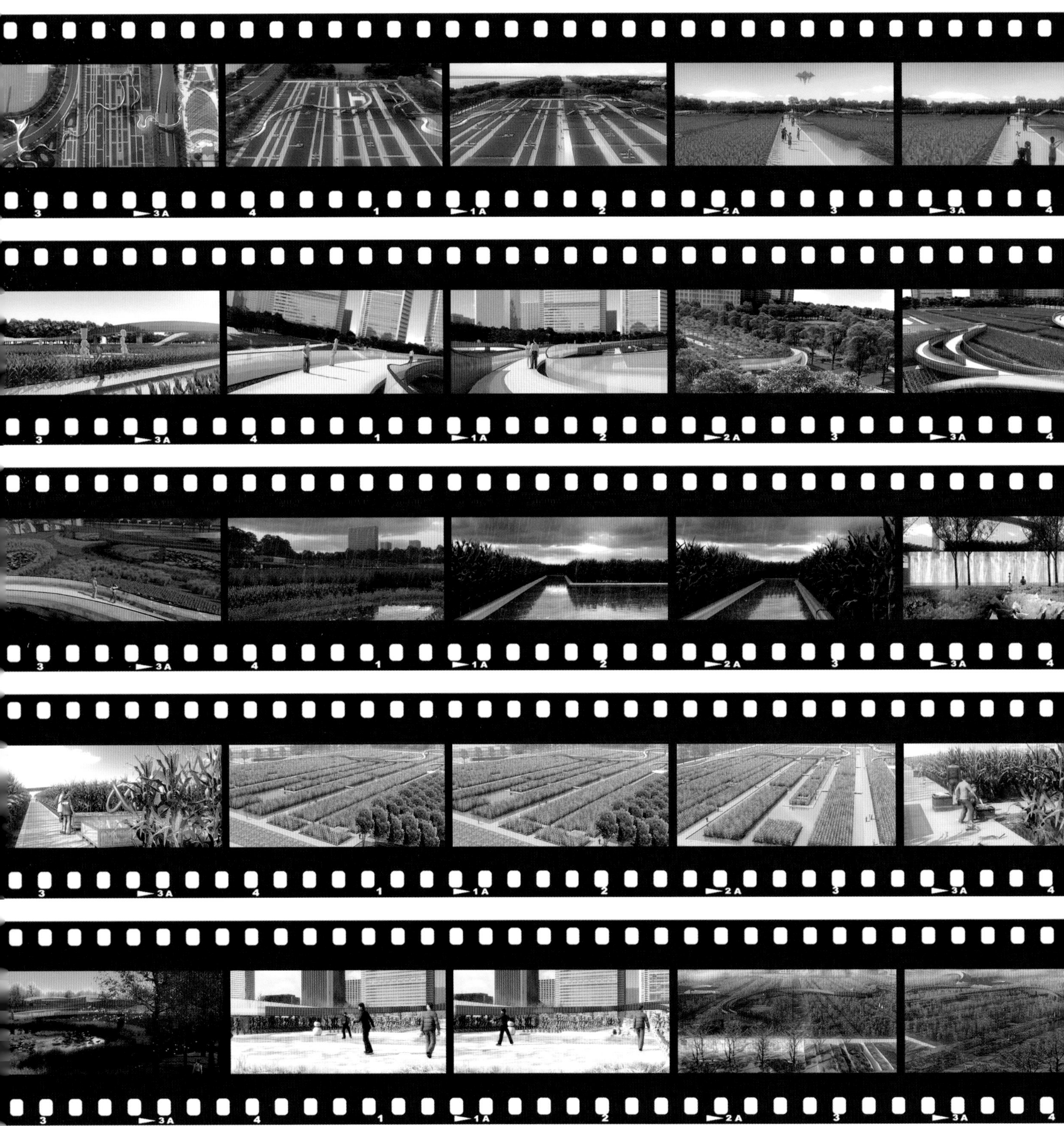

项目名称：五里界生态城暨中国光谷伊托邦城市设计
项目地点：武汉市 五里界
动画时长：5分30秒
委 托 方：湖北大都置业有限公司

Project name: Wuli Eco-city and Urban Design of China Optics Valley
Location: Wuli, Wuhan City
Movie Length: 5min 30sec
Client: Hubei Metro Properties Limited

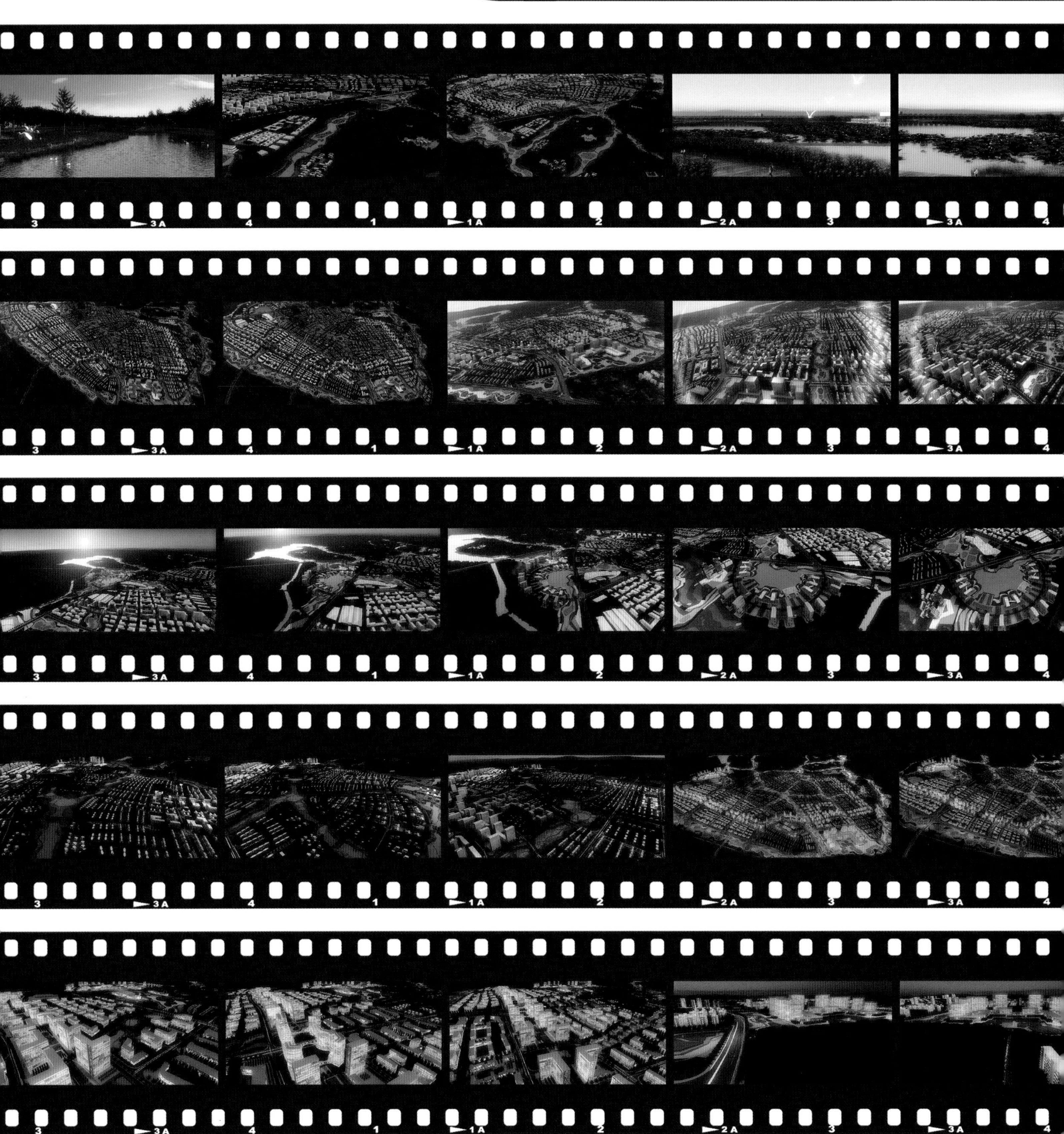

项目名称：六盘水荷城古镇改造及周边用地修建性详细规划

项目地点：贵州省 六盘水市

动画时长：2分30秒

委 托 方：六盘水经济开发区管委会开发总公司

Project name: Redevelopment of Old Dutch Town, Liupanshui CBD, and rezoning and masterplanning of surrounding area

Location: Liupanshui City, Guizhou Province

Movie Length: 2min 30sec

Client: Liupanshui Economic Development Zone Development Corporation

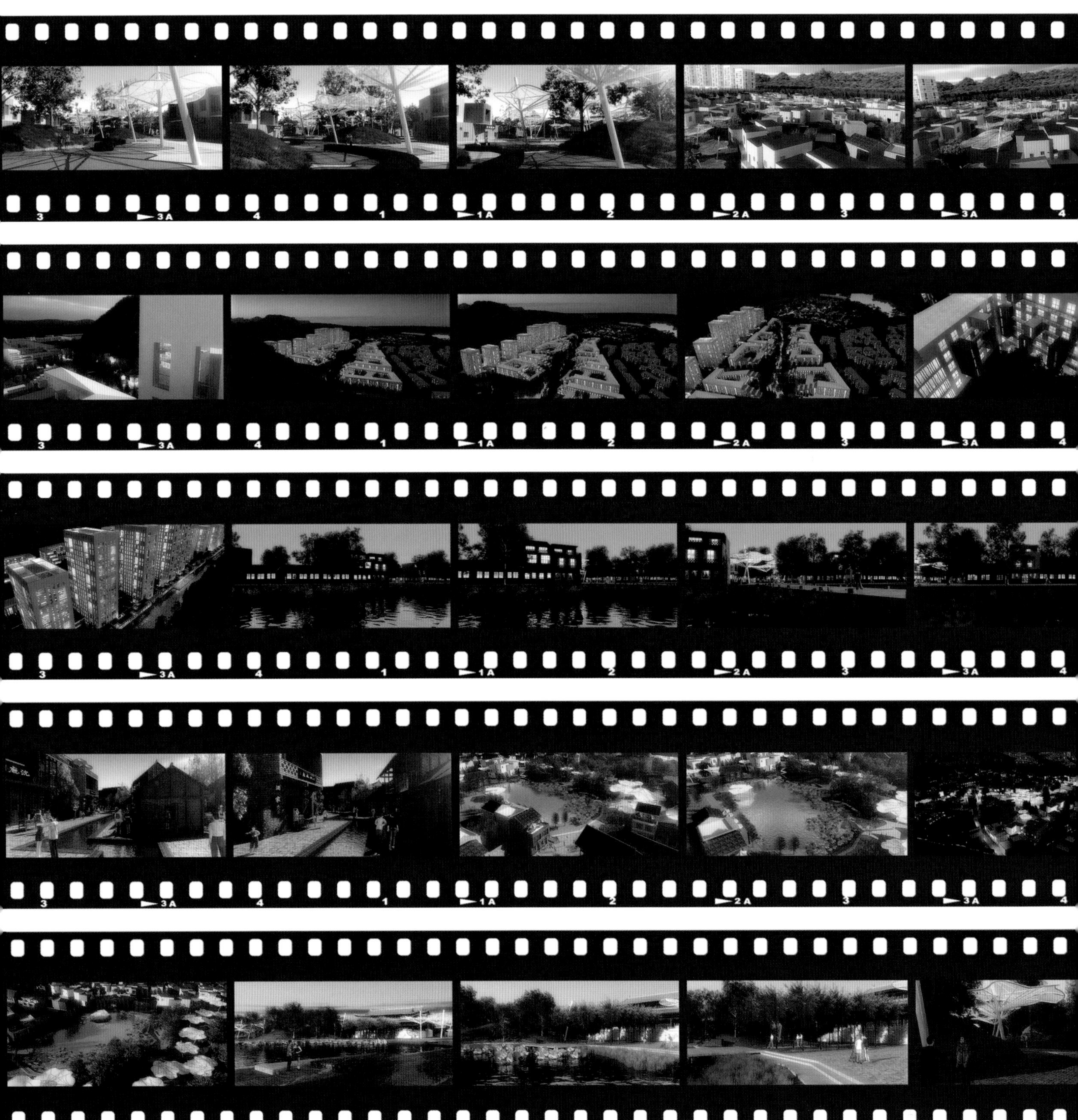

项目名称：台州鉴洋湖湿地公园一期修建性详细规划
项目地点：浙江省 台州市 黄岩区
动画时长：6分钟
委 托 方：黄岩区鉴洋湖湿地保护开发指挥部

Project name: Kamyang Lake Wetland Park – Stage 1 Masterplan
Location: Huangyan district, Taizhou City, Zhejiang Province
Movie Length: 6min
Client: Kamyang Lake Wetland Conservation and Development Headquarters, Huangyan District

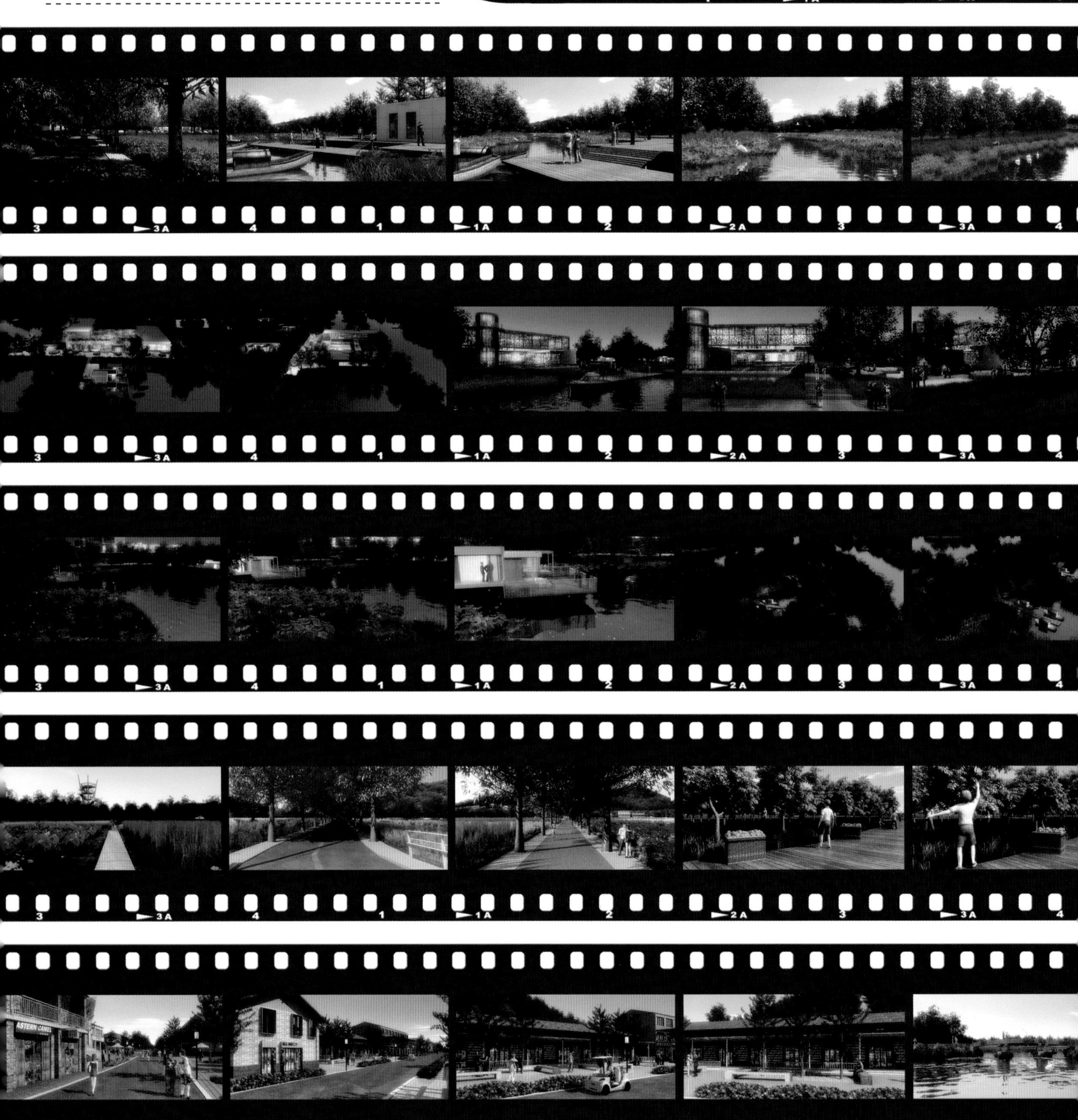

项目名称：北京市雁栖湖生态发展示范区整体规划
项目地点：北京市 怀柔区
动画时长：4分30秒
委 托 方：北京市规划委员会

Project name: Master plan of Yanqi Lake ecological demonstration region, Beijing
Location: Huairou district, Beijing
Movie Length: 4min 30sec
Client: Beijing Planning Committee

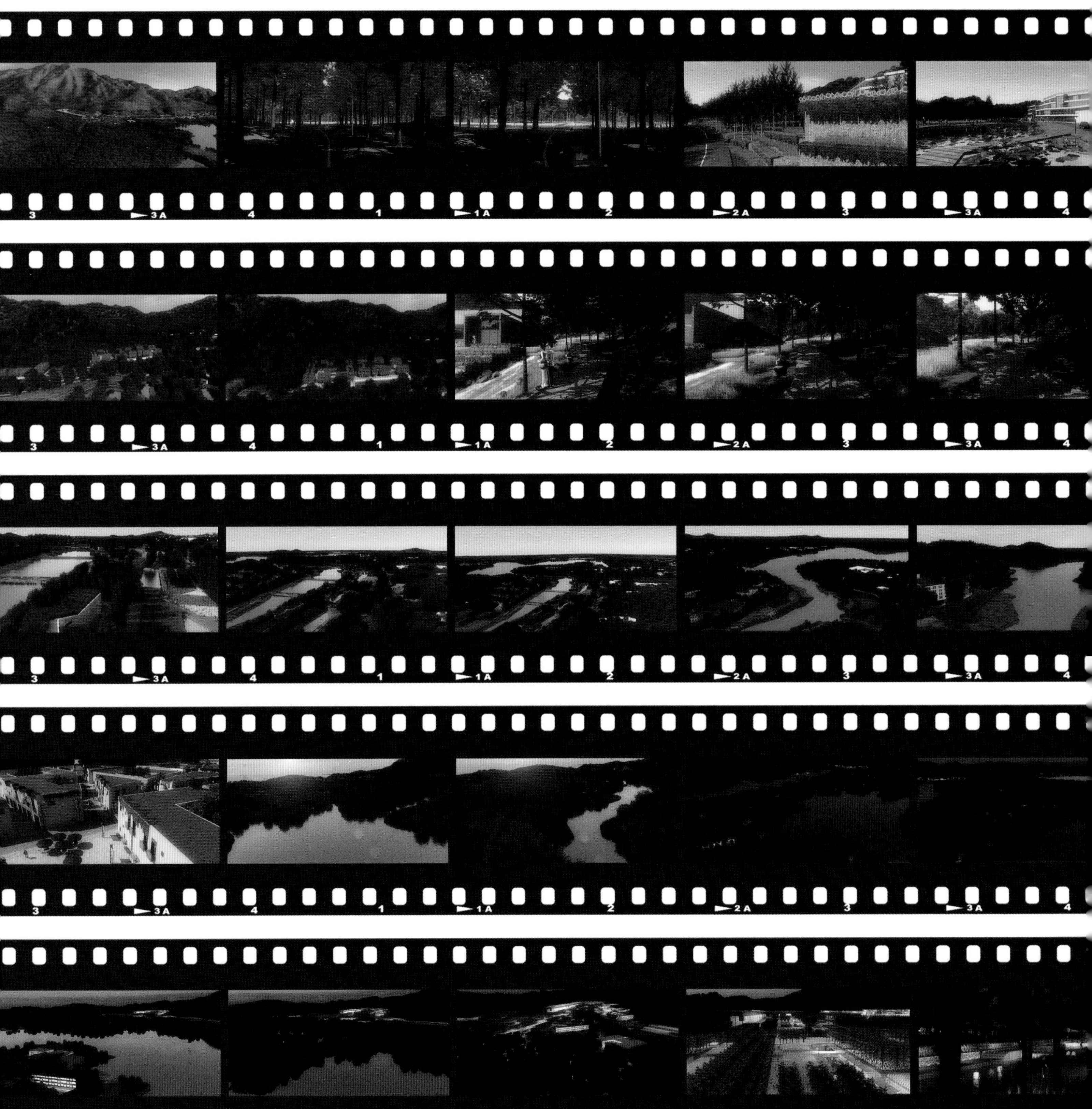

项目名称：台儿庄古镇景观规划暨核心区城市设计
项目地点：山东省 枣庄市
动画时长：5分钟
委 托 方：枣庄市城市规划局

Project name: Landscape planning and urban Design of central Taierzhuang Ancient Town
Location: Zaozhuang City, Shandong Province
Movie Length: 5min
Client: Zaozhuang City Planning Bureau

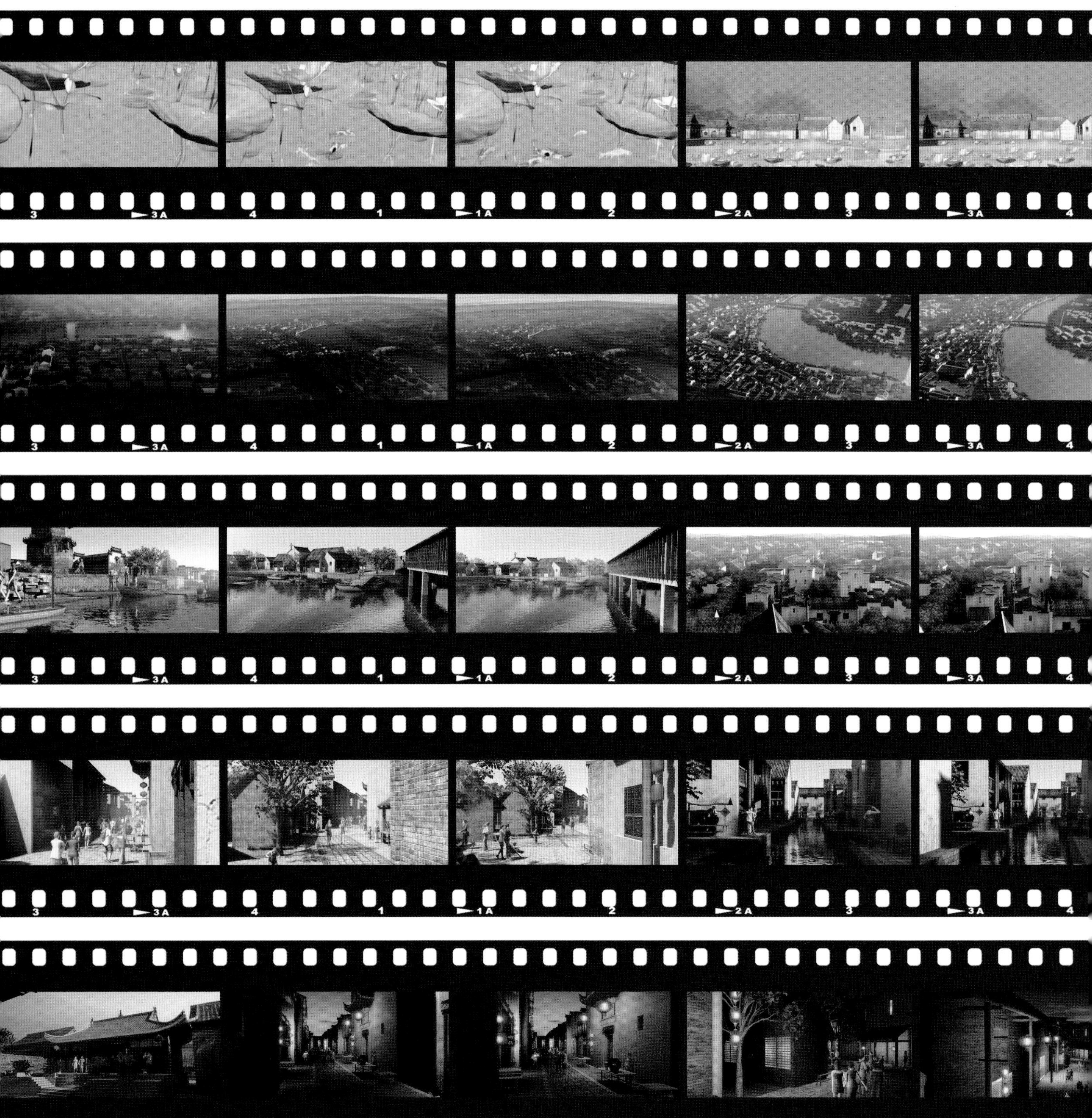

项目名称：云南祥云“新云南驿”项目概念性总体规划
项目地点：云南省 大理市 祥云
动画时长：11分40秒
委 托 方：云南阳光集团

Project name: Conceptual master planning of “New Yunnan Courier Station”, Xiangyun, Yunnan province
Location: Xiangyun, Dali city, Yunnan province
Movie Length: 11min 40sec
Client: Yunnan Sunshine Group

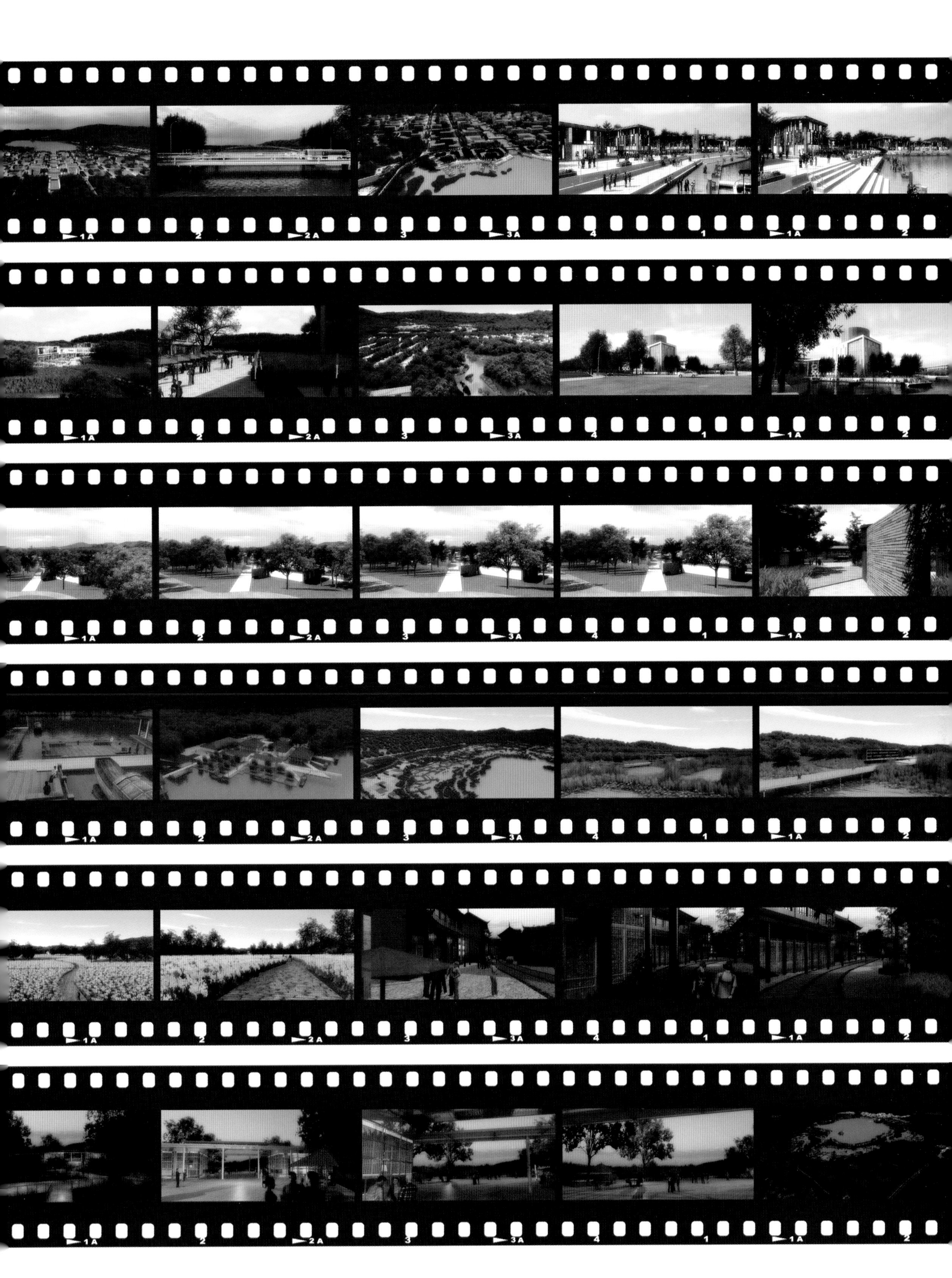

项目名称：吉安市江子头村改造规划及建筑和景观概念设计方案
项目地点：江西省 吉安市
动画时长：3分30秒
委 托 方：吉安市规划局

Project name: Reconstruction Planning and Conceptual architectural and landscape planning of Jiangzitou Village, Ji'an
Location: Ji'an City, Jiangxi Province
Movie Length: 3min 30sec
Client: Ji'an city planning bureau

项目名称：山西省临汾市洪洞县滨河西区城市设计
项目地点：山西省 临汾市 洪洞县
动画时长：6分钟
委 托 方：洪洞县住建局

Project name: Urban Design of West Riverside of Hongdong County, Linfen, Shanxi
Location: Hongdong County, Linfen City, Shanxi Province
Movie Length: 6min
Client: Hongdong Housing and urban-rural construction Bureau

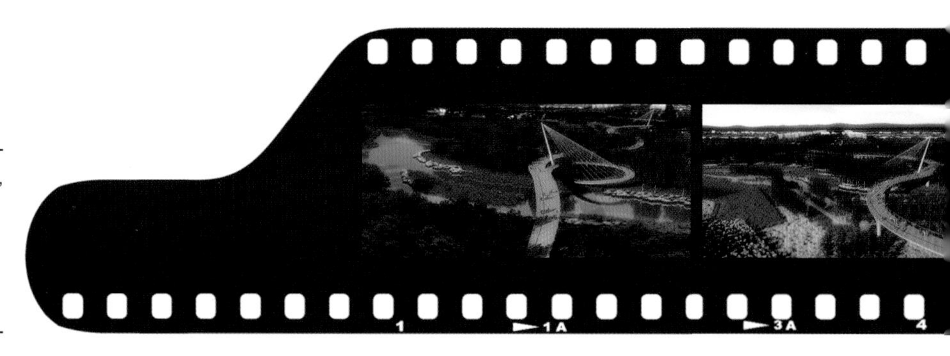

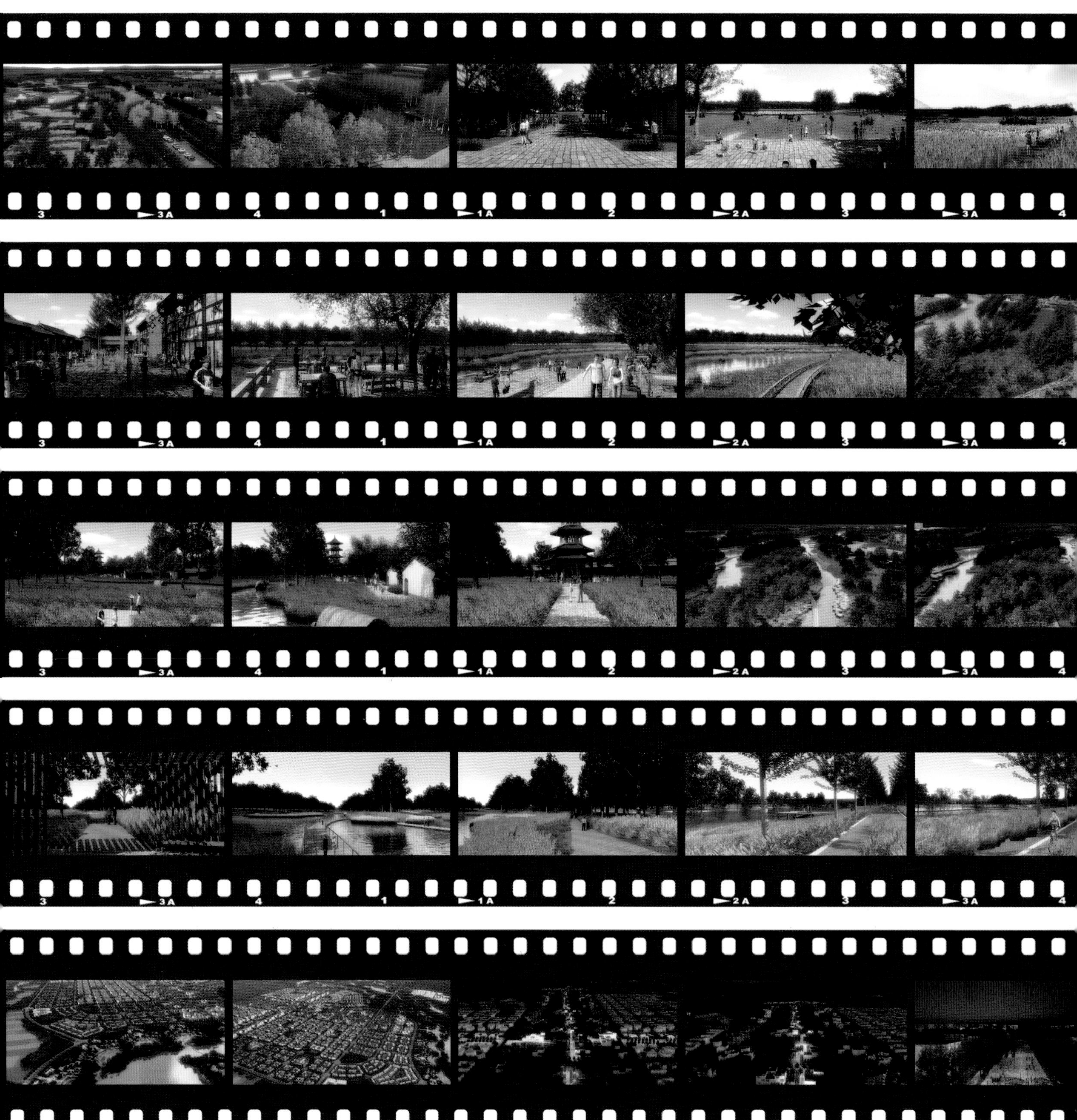

项目名称：中国滑县道口古镇
项目地点：河南省 安阳市 滑县
动画时长：3分30秒
委 托 方：滑县城乡规划管理局

Project name: Daokou Ancient Town, Hua County
Location: Hua County, Anyang City, He'nan Province
Movie Length: 3min 30sec
Client: Hua County Urban and Rural Housing Authority

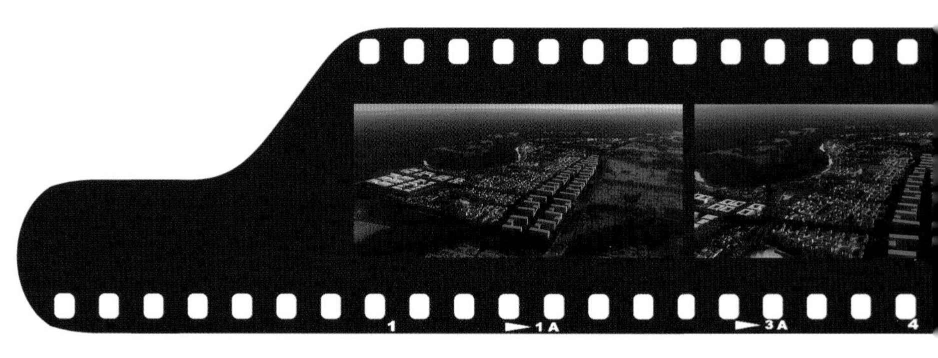

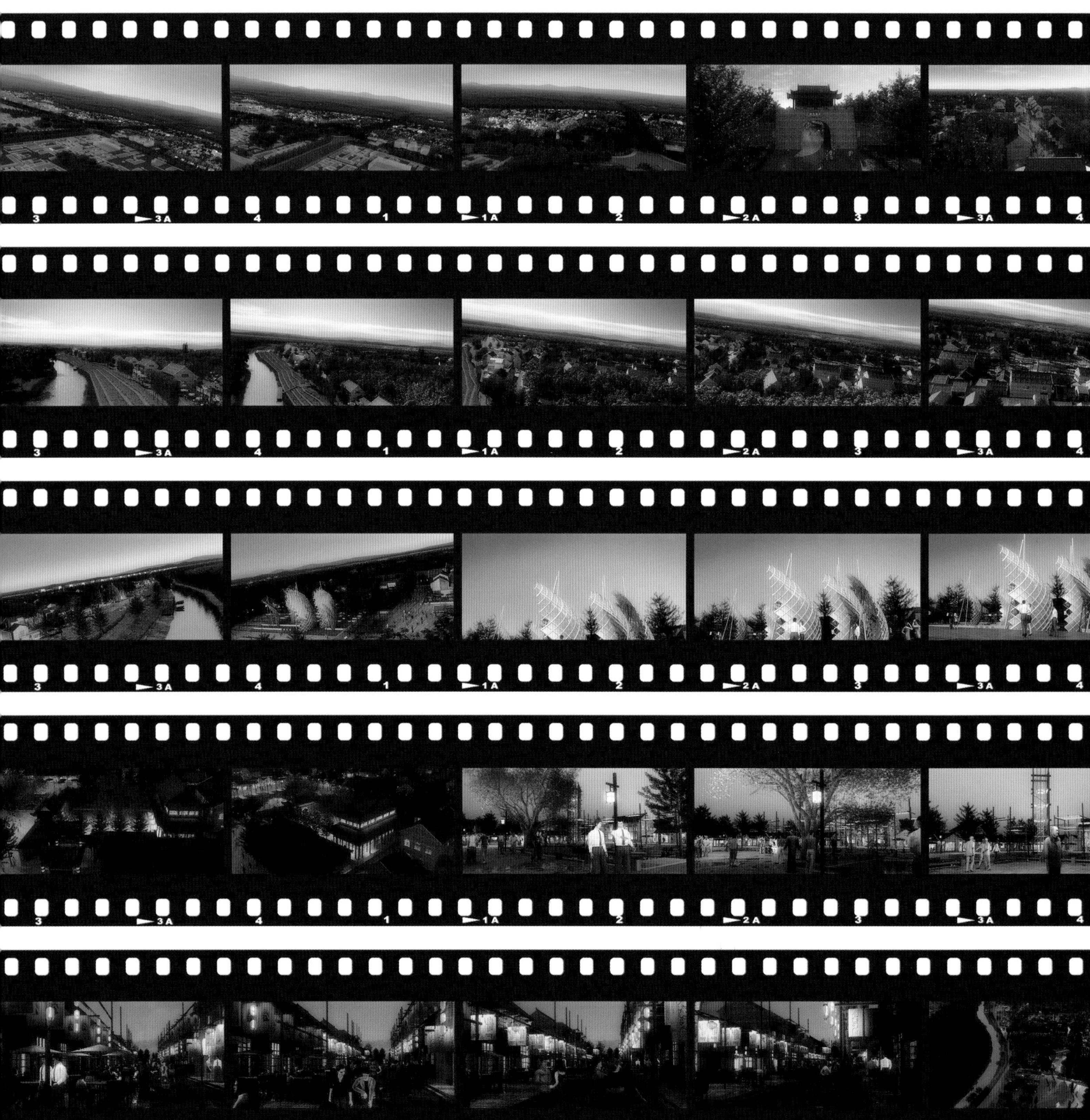

项目名称：郑州IT产业园概念规划
项目地点：河南省 郑州市
动画时长：4分钟
委 托 方：郑州国家高新技术产业开发管委会

Project name: Conceptual Planning of Zhengzhou IT Industrial Park
Location: Zhengzhou City, He'nan Province
Movie Length: 4min
Client: Zhengzhou National Industrial Development Zone Authority

项目名称：钟祥市明显陵风景区
项目地点：湖北省 钟祥市
动画时长：7分30秒
委 托 方：钟祥市人民政府

Project name: The Ming Tombs Scenic Area, Zhongxiang
Location: Zhongxiang City, Hubei Province
Movie Length: 7min 30sec
Client: Zhongxiang People's Government

项目名称：重庆市江津区新城城市设计
项目地点：重庆市 江津区
动画时长：3分钟
委 托 方：重庆市规划研究中心

Project name: Urban Design of Jiangjin district of Chongqing City
Location: Jangjin district, Chongqing City
Movie Length: 3min
Client: Chongqing Planning Research Centre

重庆市江津区新城城市设计

项目名称：宜昌廊桥水岸
项目地点：湖北省 宜昌市
动画时长：10分钟
委 托 方：湖北大都置业有限公司

Project name: Yichang Bridge and Water Residential Area
Location: Yichang City, Hubei Province
Movie Length: 10min
Client: Hubei Dadu Investment Co., Ltd.

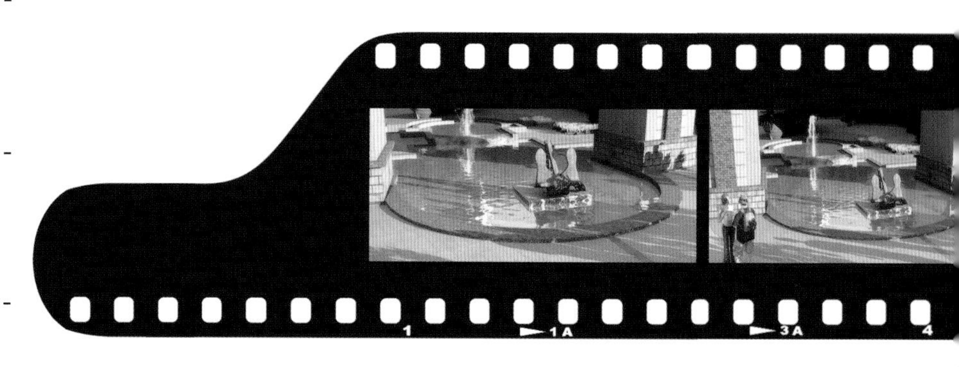

项目名称：济宁阳光城市花园
项目地点：山东省 济宁市
动画时长：14分30秒
委 托 方：山推置业

Project name: Sunshine Urban Garden Residential Area, Jining
Location: Jining City, Shandong Province
Movie Length: 14min 30sec
Client: Shantui Investment Co., Ltd.

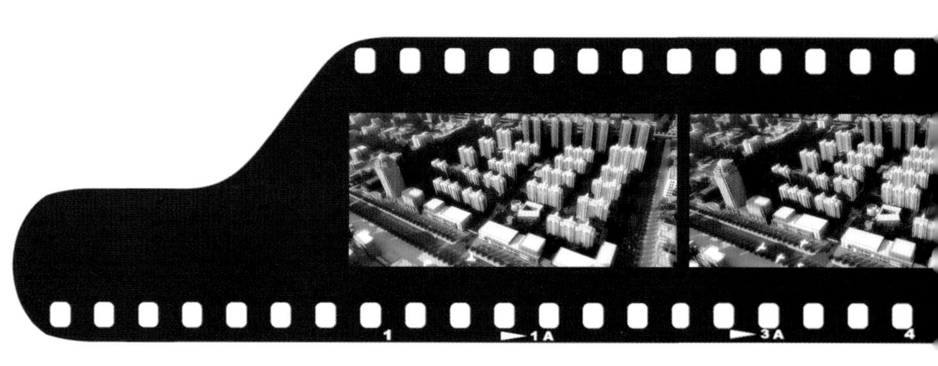

1A 2 2A 3 3A 4 1 1A 2
1A 2 2A 3 3A 4 1 1A 2
1A 2 2A 3 3A 4 1 1A 2
1A 2 2A 3 3A 4 1 1A 2
1A 2 2A 3 3A 4 1 1A 2
1A 2 2A 3 3A 4 1 1A 2

项目名称：京林理想城
项目地点：河南省 林州市
动画时长：4分钟
委 托 方：河南中房威泰置业有限公司

Project name: Jinglin Ideal City
Location: Linzhou City, He'nan Province
Movie Length: 4min
Client: Henan Zhong Fang Wei Tai Investment Co., Ltd.

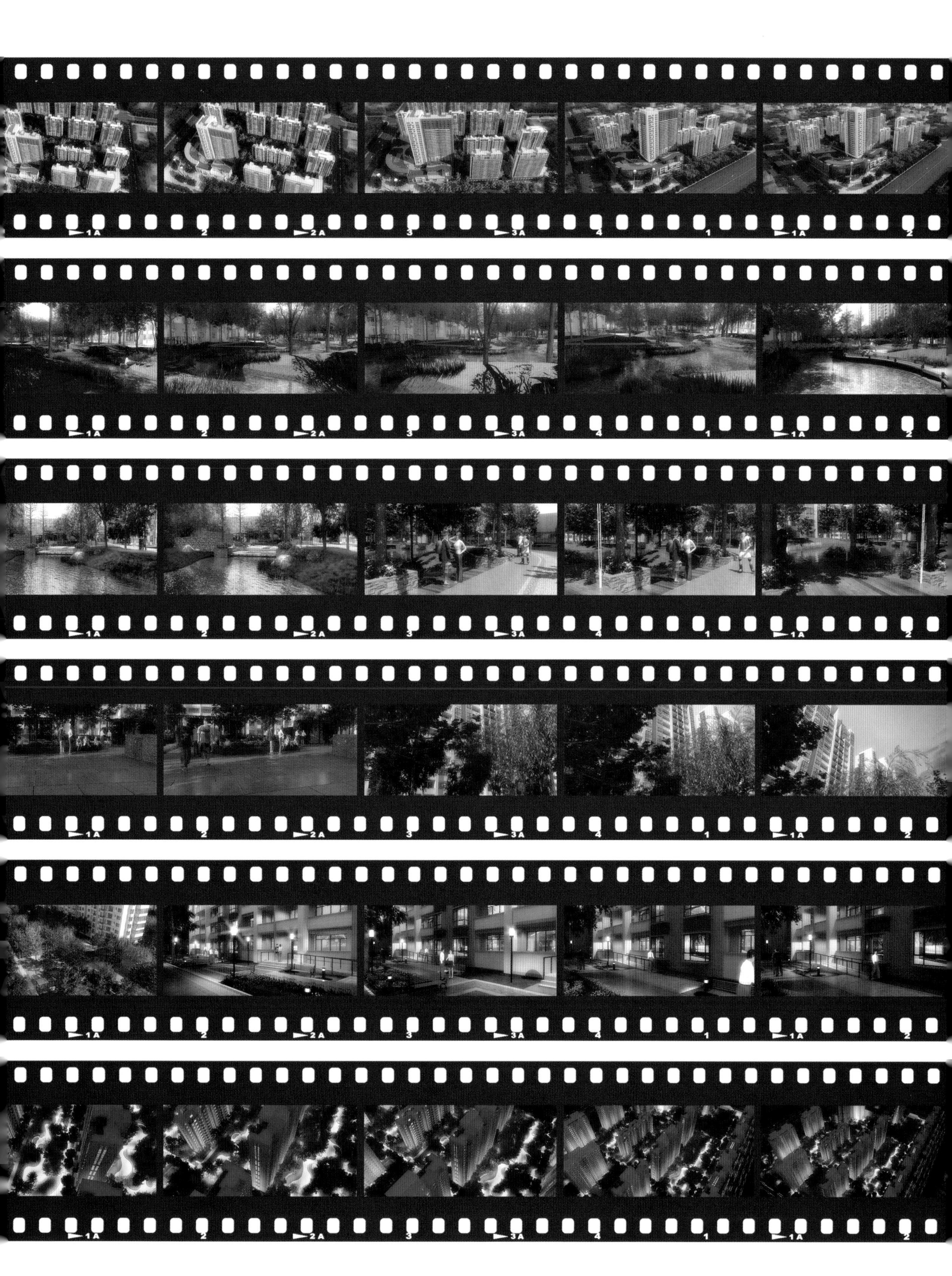
1A 2 2A 3 3A 4 1 1A 2
1A 2 2A 3 3A 4 1 1A 2
1A 2 2A 3 3A 4 1 1A 2
1A 2 2A 3 3A 4 1 1A 2
1A 2 2A 3 3A 4 1 1A 2
1A 2 2A 3 3A 4 1 1A 2

项目名称：海拉尔河东新城核心区城市设计
项目地点：内蒙古 呼伦贝尔市
动画时长：5分30秒
委 托 方：内蒙古蒙西建设集团有限公司

Project name: Urban Design of Central Hailaer East New Town
Location: Hulunbeier, Inner Mongolia
Movie Length: 5min 30sec
Client: Mongxi Investment Group

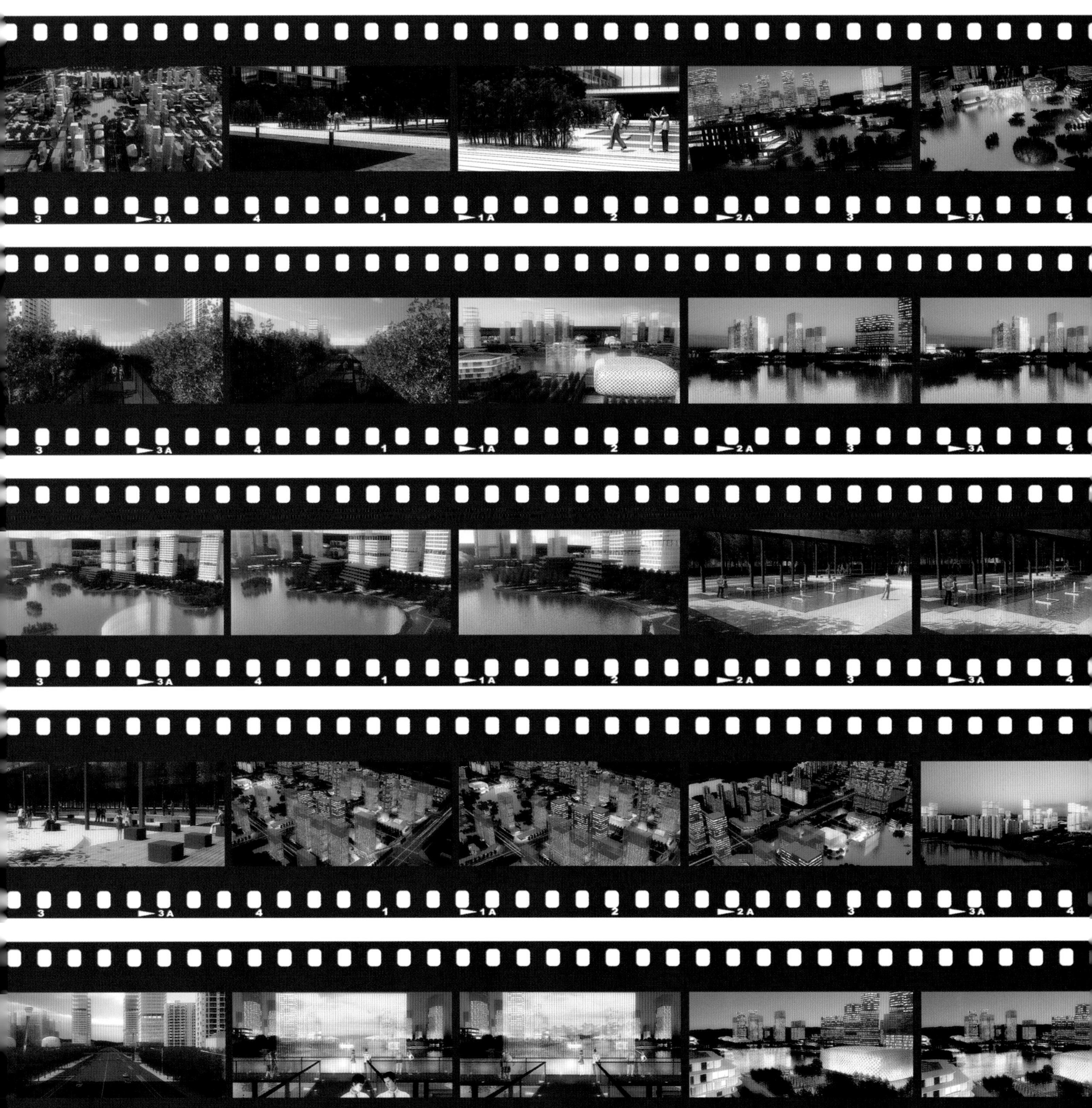

项目名称：胶南市风河两岸景观工程
项目地点：山东省 胶南市
动画时长：4分钟
委 托 方：胶南市建设局

Project name: Landscape Project Feng He Riverside, Jiaonan
Location: Jiaonan City, Shandong Province
Movie Length: 4min
Client: Jiaonan Construction Bureau

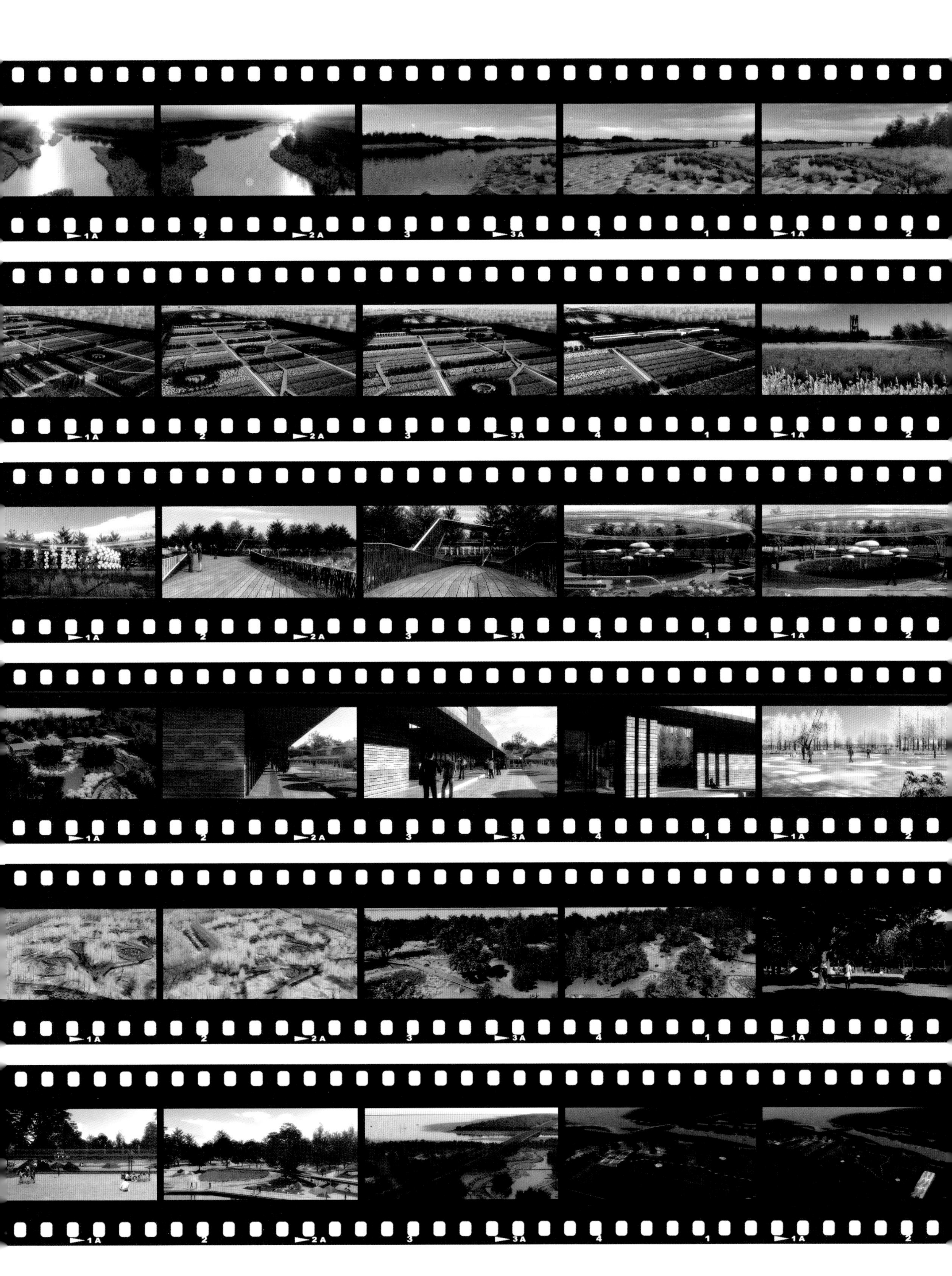
1A 2 2A 3 3A 4 1 1A 2
1A 2 2A 3 3A 4 1 1A 2
1A 2 2A 3 3A 4 1 1A 2
1A 2 2A 3 3A 4 1 1A 2
1A 2 2A 3 3A 4 1 1A 2
1A 2 2A 3 3A 4 1 1A 2

项目名称：山东海阳龙栖城修建性概念规划
项目地点：山东省 海阳市
动画时长：3分钟
委 托 方：山东烟台东方海洋置业有限公司

Project name: Construction Conceptual Planning of Longqi City, Haiyang, Shandong
Location: Haiyang City, Shandong Province
Movie Length: 3min
Client: Shandong Yantai Dong Fang Hai Yang Investment Co., Ltd.

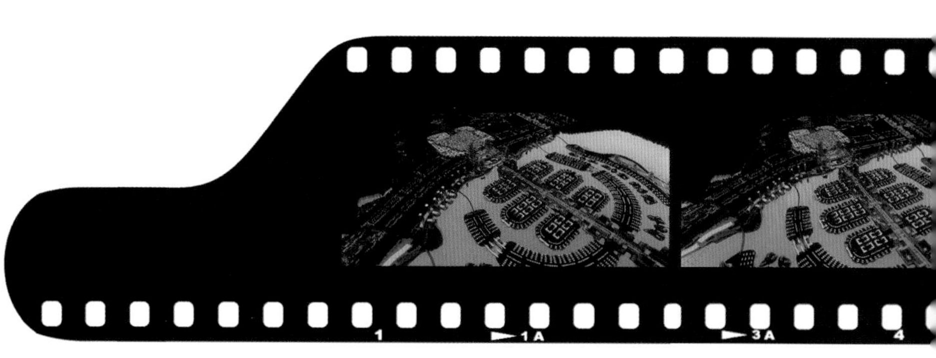

项目名称：洛阳师范学院校园总体规划方案
项目地点：河南省 洛阳市 伊滨区
动画时长：4分钟
委 托 方：洛阳师范大学

Project name: Master plan of Luoyang Normal University Campus
Location: Yibin, Luoyang City, He'nan Province
Movie Length: 4min
Client: Luoyang Normal University

部分获奖项目
Some Awards Received

□ 迁安三里河生态廊道
2011 世界建筑节全球最佳景观奖
The Qian'an Sanlihe Greenway
2010 World Architecture Festival Awards, World's Best Landscape

□ 上海后滩公园
2010 美国 ASLA 卓越设计奖
2010 世界建筑节全球最佳景观奖
2011 美国建筑奖
Shanghai Houtan Park
2010 ASLA Excellence Design Award
2010 World Architecture Festival Awards,World's Best Landscape
2011 American Architecture Awards

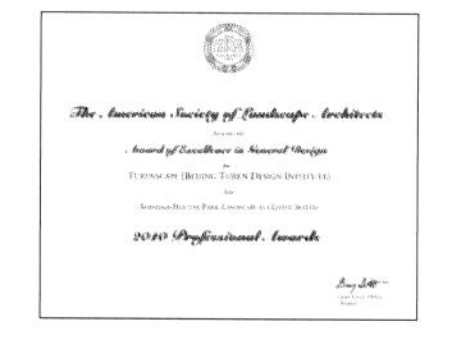

□ 天津桥园公园
2010 美国 ASLA 荣誉设计奖
2009 世界建筑节全球最佳景观奖
2009 中国人居环境范例奖
Tianjin Qiaoyuan Park
2010 ASLA Honor Design Award
2009 World Architecture Festival Awards,World's Best Landscape
2009 Human Habitat Award of China

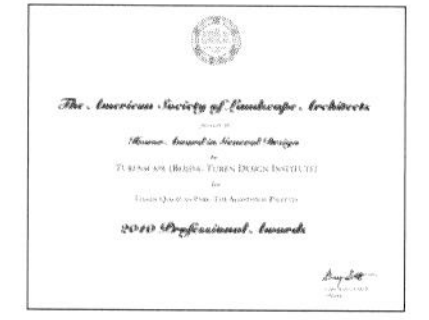

□ 秦皇岛滨海景观带
2010 美国 ASLA 荣誉设计奖
2009 世界滨水设计最高荣誉奖
The Qinhuangdao Beach Restoration
2010 ASLA Honor Design Award
2008 Excellence on the Waterfront Top Honor Award

□ 秦皇岛汤河公园
2007 美国 ASLA 荣誉设计奖
2008 世界建筑奖
2008 首届国际建筑展览会优秀奖
The Red Ribbon Park,Qinghuangdao
2007 ASLA Honor Design Award
2008 The International Architecture Award
2008 Highly Commended,The Inaugural World Architecture Festival

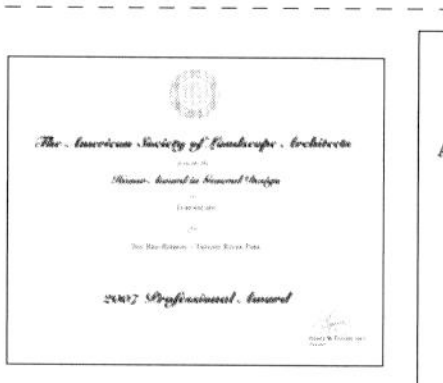

□ 中山岐江公园
2002 美国 ASLA 荣誉设计奖
2008 世界滨水设计最高荣誉奖
2009 ULI 全球杰出奖，亚太杰出奖
Zhongshan Shipyard Park
2002 ASLA Honor Design Award
2008 Excellence on the Waterfront Top Honor Award, Excellence Award for the Asian Pacific Region
2009 ULI Global Awards For Excellence

□ 黄岩永宁公园
2006 美国 ASLA 专业设计奖
2007 世界滨水设计最高荣誉奖
2006 中国人居环境范例奖
Yongning River Park, Huangyan
2006 ASLA Design Honor Award
2007 Excellence on the Waterfront Top Honor Award
2006 Human Habitat Award of China

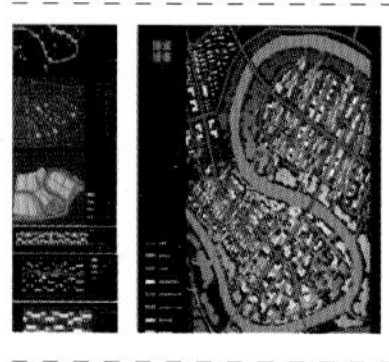

□ “反规划”– 台州案例
2005 美国景观设计师协会荣誉规划奖
Taizhou City, Negative Approach to Planning
2005 ASLA Honor Award

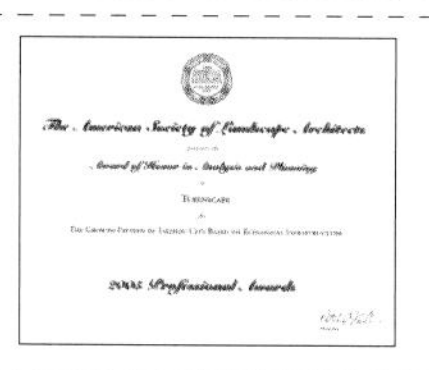

□ 沈阳建筑大学
2005 美国 ASLA 荣誉设计奖
2007 世界青年建筑师奖
2009 中国人居环境范例奖
Shenyang Jianzhu University Campus
2005 ASLA Professional Awards
2007 Award Commended (The Architecture Review)
2009 Human Habitat Award of China

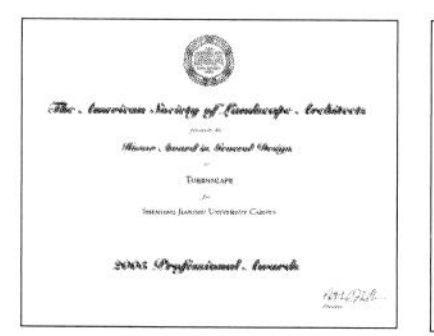

□ 都江堰广场
2005 世界青年建筑师奖
Dujiangyan Square
2005 Award Commended (The Architecture Review)

图书在版编目（CIP）数据

土人景观最新规划设计表现图集 / 北京土人景观与建筑规划设计研究院编 .
-- 北京 : 中国林业出版社 , 2012.6
ISBN 978-7-5038-6512-1

Ⅰ . ①土… Ⅱ . ①北… Ⅲ . ①景观设计 - 作品集 - 中国 - 现代
Ⅳ . ① TU986.2

中国版本图书馆 CIP 数据核字 (2012) 第 040456 号

参与制图人员名单：

邢春杰 班明辉 李媛媛 李 良 滕宝志 曹 伟 彭怡博 栾 杨
苗 禹 刘 超 刘术铭 强鹏飞 杨 君 霍庆海 彭 珍 刘兆其
徐朗峰 孙传杰 陈 枫 凌 宏 陈 波 王纪周 曹延春 欧阳煜
冷 刚 李东泽 王 策 张云波 翟 君 王 亮 孟雨嘉 栾冠军
董星月 李德鑫 法云飞 李海天

整体设计：北京湛和文化发展有限公司
http://www.anedesign.com

中国林业出版社 • 建筑与家居图书出版中心

责任编辑：纪 亮 / 李 顺
出版咨询：（010）83223051

出 版：中国林业出版社（100009 北京西城区德内大街刘海胡同 7 号）
China Forestry Publishing House
网 站：http://lycb.forestry.gov.cn/
印 刷：北京利丰雅高长城印刷有限公司
发 行：新华书店北京发行所
电 话：（010）83224477
版 次：2012 年 6 月第 1 版
印 次：2012 年 6 月第 1 次
开 本：889mm × 1194mm 1/16
印 张：16.5
字 数：100 千字
定 价：288.00 元